AF820810

Introduction to Plasma Physics

Introduction to Plasma Physics

Martin A. Uman

Associate Professor
of Electrical Engineering
University of Arizona
Tucson, Arizona

McGraw-Hill Book Company

New York
San Francisco
Toronto
London

Introduction to Plasma Physics

Library of Congress Catalog Card Number 63-21546

3 4 5 6 7 8 9 10 – MP – 1 0 9 8 7

65744

To

Dorit Brigitta

and

Mara Anne

Preface

"Introduction to Plasma Physics" is designed to be used as a textbook for a one-semester introductory course in plasma fundamentals. It is written for engineers at the senior or first-year-graduate level, but it should also be useful to those engineers and physicists in industry who wish to acquire a basic understanding of plasma phenomena. The primary objective of "Introduction to Plasma Physics" is to introduce the reader to the physical and mathematical concepts which are fundamental to all branches of plasma physics and engineering. A knowledge of the fundamentals of plasma physics will enable the reader to understand, in a unified context, portions of such superficially diverse areas of study as radio propagation, astrophysics, magnetohydrodynamic power generation, microwave amplification, space-vehicle propulsion, controlled-thermonuclear-fusion processes, and gas discharge devices. After studying "Introduction to Plasma Physics," the reader will be prepared to undertake further reading at an advanced level in any of the areas of plasma physics.

An attempt has been made to develop the fundamentals of plasma physics in a manner which does not require a formal background in physics. While many engineering students do have a working knowledge of kinetic theory, statistical mechanics, fluid dynamics, and thermodynamics, most are deficient in these important areas. It is therefore in keeping with the objectives of the text that the needed elements of these subjects be developed within it. Thus a certain amount of material has been included which, from the viewpoint of a graduate student in physics, would be termed "elementary." It has been the author's experience in teaching plasma physics to first-year-graduate-level electrical engineers at the University of Arizona that this material is not elementary to them. The expert in plasma physics may find that the inclusion of this material lends an uneven quality to certain portions of the text. It is, however, the lack of this material which makes several other plasma books very difficult for the engineering student to follow.

The reader of "Introduction to Plasma Physics" is assumed to have an undergraduate knowledge of electromagnetic field theory, vector analysis, and differential equations. It is helpful for the reader to have had some exposure to kinetic theory, statistical mechanics, fluid dynamics, thermodynamics, and tensor analysis, although such exposure is not mandatory since the needed elements of these subjects are developed within the text. In general, senior standing in the sciences or in engineering is a prerequisite to the use of the book.

The contents of the text can be adequately covered in a one-semester course. The material in Appendix A, The Derivation of the Momentum Transfer Equation and the Continuity Equation from the Boltzmann Equation, and in Appendix B, The Frictional-force Term and the Effective Collision Frequency, may be included in Chap. 4 or may be excluded entirely, at the discretion of the instructor. The assumed form of the collision term used in the momentum transfer equation (introduced in Chap. 4) is not the conventional form for that term. The form assumed, however, provides an adequate description of those collision-influenced plasma phenomena considered throughout the text. The more conventional approach to the inclusion of the effects of collisions in the momentum transfer equation is given in Appendix B. The material in Appendix D should be included in Chap. 5 if the reader is unfamiliar with elementary tensor operations.

The problems are an important part of the text. All problems should be solved either by the student or by the instructor as part of the lecture.

Rationalized mks units are used throughout. Occasionally an electron volt or an angstrom may appear, but such occurrences are well documented.

The author wishes to express his gratitude to the following gentlemen who have made significant contributions to "Introduction to Plasma Physics" by reading and criticizing the text while it was in preliminary form: Professor P. A. Sturrock, Professor A. W. Trivelpiece, and Mr. Myron F. Uman. Credit is also due to Mrs. Betty Perkins and Mrs. Yvonne Walrath, who typed the several preliminary versions of the manuscript, and to Miss Rosemarie Stampfel, who typed the final draft.

Martin A. Uman

Contents

chapter 4 The Equations of Plasma Physics 26

chapter 5 The Behavior of Charged Particles in Electric and Magnetic Fields 41

chapter

1

What Is a Plasma?

A plasma may be defined as a conglomeration of mobile positively and negatively charged particles, there being approximately the same density of positive charges as of negative. In point of fact it is not necessary for both groups of charge carriers to be mobile for the system to display plasma characteristics; one group can be stationary, as, for example, are the positive lattice ions of a metallic solid. For a plasma to display plasma characteristics, however, the average distances between the charged plasma particles must be very much smaller than the physical dimensions of the total plasma. The plasma system can also contain neutral particles and electromagnetic radiation (photons). In order to define a plasma more precisely (or, in fact, to define anything precisely) one must resort to mathematics. When we have become familiar with the mathematics of plasmas, we shall be able to improve our "physical" definition.

There are a multitude of systems which will satisfy our physical definition and which we, with some assurance, can expect to exhibit plasma characteristics. Plasmas come, broadly speaking, in solids, liquids, and gases. The metallic solid with its fixed positive charges (lattice ions) and its sea of mobile electrons can exhibit plasma characteristics. When table salt is dissolved in water it forms a plasma composed of equal numbers of mobile positive sodium and negative chlorine ions. We shall, however, be primarily concerned in this book with the gaseous plasma. A good deal of what is said about the gaseous plasma will nevertheless apply to solid and liquid plasmas.

What are the plasma characteristics that we have referred to in the preceding paragraphs? The plasma is unique among the classical states of matter in the ways that it interacts internally with itself and with externally applied fields and forces. The plasma state is capable of supporting numerous modes of spontaneous (thermally induced) and externally induced oscillations and instabilities. These oscillations and instabilities take quite peculiar forms because of (1) the way in which the plasma strives to maintain internal charge neutrality and (2) the way in which plasma particles with a range of velocities interact among themselves.

It is estimated that more than 99.9 per cent of the matter in the known universe is plasma. Interstellar space is the grandest of all imaginable plasmas. The stars that speckle this vast ocean of what was once thought to be nothingness are themselves largely in the plasma state. On our own sun one may observe quite spectacular plasma phenomena in the form of solar flares, solar prominences, and sunspots. Closer to home, the aurora borealis provides a rather impressive plasma display in the earth's upper atmosphere. The lightning which accompanies thunderstorms is a very vivid, if somewhat mathematically intractable, example of a plasma.

Since the plasma state appears in such abundance in the universe, it is only fitting and natural that man should strive to understand and benefit from the plasma. At one time, roughly before man attempted in earnest to leave the planet earth, the plasma was not too important in a practical sense. At that time, plasma research was primarily divided between those physicists and engineers who studied gas discharge phenomena and those astronomers who examined theoretically the dynamics of stellar and interstellar plasmas. Since the era of the earth satellite, the interest in plasma research and the number of people engaged in that research have grown to very great proportions. It is not possible to escape the significance of this new plasma research, as we shall see in the following discussion.

For convenience we shall divide all plasma research into five parts. The five areas, rather arbitrarily chosen and not necessarily presented in order of magnitude or importance, are (1) communication—the interaction of electromagnetic waves with plasmas, (2) controlled thermonuclear reaction—the containment of plasmas, (3) plasma-filled tubes—the exploitation of induced plasma oscillations, (4) electric power generation—the transformation of plasma motion to electricity, and (5) propulsion—the transformation of electricity to plasma motion. In the course of this book we shall encounter fundamental concepts which are used in all of these areas of plasma research.

1. The word "communication" implies the transfer of information from one point to another. A very convenient way to transfer information is to allow the information to hitchhike a ride on an electromagnetic wave. The electromagnetic wave travels through particle-free space (vacuum) at the rather rapid speed of about 3×10^8 m/sec. If the medium through which the wave travels is populated by charged particles, the properties of the wave may be radically altered. The information carried by the electromagnetic wave may be changed or destroyed. The plasma can be very destructive. It is interesting to note that if one knows all about the electromagnetic wave that has been sent and can compare this transmitted wave with the received plasma-distorted wave, a good deal of information about the plasma can be determined.

Radiation leaving an antenna need not travel far before encountering a natural plasma. The earth's atmosphere from an altitude of approximately 70 to over 300 km is a plasma. This natural blanket of plasma, the ionosphere, has long been used to reflect radio waves back to earth. The particles of the ionosphere, incidentally, are ionized by radiation from the sun. Further out into space there are zones of plasma trapped in the earth's magnetic field; the particles which are trapped in these zones are supplied by the sun. And, of course, man can be counted on to create a plasma whenever it is detrimental to communication processes. A good example of such a man-made plasma is the shock-wave-induced plasma that surrounds the body of a space vehicle during high-speed entry of the earth's atmosphere. Communication through this plasma layer cannot be effected by straightforward means.

Plasmas are capable of doing very strange things to electromagnetic waves. One electromagnetic wave traveling through a plasma may, under certain circumstances, interact with a second wave in such a way that the modulation or information imposed on the first wave is transferred to the second. Cross modulation of this type requires a nonlinear medium, and the plasma can be just such a medium. Radio Luxembourg had the honor of first producing a measured plasma-induced cross modulation, and hence the cross modulation is called the "Luxembourg effect." Because a plasma exhibits nonlinear properties, it has many of the same applications (particularly in wave guides) as nonlinear solid-state materials.

At low radio frequencies, in the presence of a magnetic field, electromagnetic waves may propagate through a plasma with a velocity which can be many orders of magnitude less than the speed of light. These low-frequency low-velocity electromagnetic waves are called "magnetohydrodynamic" or "hydromagnetic" waves. Hydromagnetic waves have been observed in the laboratory and have been induced in the upper atmosphere by bomb blast.

2. A hydrogen bomb, when exploded, is a thermonuclear reaction out of control. Impressive amounts of energy are released in a very short time. If the thermonuclear reaction were slowed down, that is, controlled, the output energy could be useful to man. The thermonuclear reactions of interest to us are the combinations or fusions of various isotopes of hydrogen to give particle products and large amounts of energy. The process of fusion creates energy by virtue of the fact that the mass of the particle products is less than the sum of the masses of the original hydrogen isotopes. Specifically, the reactions of interest are

$$\mathrm{D} + \mathrm{D} \begin{array}{l} \nearrow \mathrm{He}^3 + n + 3.2\ \mathrm{Mev} \\ \searrow \mathrm{T} + p + 4.0\ \mathrm{Mev} \end{array}$$

$$\mathrm{D} + \mathrm{T} \rightarrow \mathrm{He}^4 + n + 17.6\ \mathrm{Mev}$$

where D represents deuterium (one proton and one neutron); T, tritium (one proton and two neutrons); He, helium; n, neutron; p, proton; and Mev, million electron volts (an electron volt is about 1.6×10^{-19} joule). Since deuterium occurs in nature, it presents an almost inexhaustible supply of fuel for the fusion process. In addition, the fusion products are nonradioactive, and the reaction is fail-safe; that is, there is no possibility of a runaway.

For the fusion of two particles to occur, the two particles must stay sufficiently close to each other long enough for the short-range nuclear forces to act on the particles and to pull them together. This is the thermonuclear reaction. In order for particles to get close enough together so that the nuclear forces may act, the repulsive Coulomb forces must be overcome. Thus the particles must have a high energy. The most practical laboratory system capable of operating at the necessary energies is the fully ionized plasma.

The fundamental problem of thermonuclear-reaction control is that of containing a high-energy plasma long enough so that an appreciable number of fusions can take place. Solid walls make an ineffective container because they cool the plasma. Gravitational forces are not of sufficient magnitude to contain a man-made plasma (although they are quite effective in holding stars together). One concludes that the plasma must be held together by electric or magnetic forces or combinations of the two. Since a static electric field causes charge separation, it can be ruled out as a candidate for the container. The remaining candidates are magnetic fields and combinations of magnetic and electric fields. In the course of this book we shall have an opportunity to consider the following schemes for the containment of a high-energy plasma: (*a*) the magnetic mirror—a static magnetic field in which charged particles are trapped much as solar particles are trapped in the earth's magnetic field, (*b*) the pinch—the self-constriction of a cylinder of plasma current due to its own magnetic field, and (*c*) the cusped or picket-fence geometry—a static magnetic field designed especially for particle containment. All containment schemes suffer from problems of plasma instabilities and oscillations. It is in part the plasma instabilities that inhibit the length of time that a plasma can be contained.

The problems associated with the interaction of plasmas both with containing fields and with themselves have established plasma physics as a discipline to be reckoned with. The behavior of groups of charged particles in electric and magnetic fields is a subject of much interest and much difficulty.

3. It may surprise the reader to find out that the fluorescent lamp under which he sits each day is a plasma device. Radiation from the gaseous discharge plasma inside the glass activates the fluorescent material with which the tube is coated. The fluorescent material then emits light. In the display-sign neon tube the direct radiation from a neon plasma is visible. Among the numerous pre-plasma-era plasma devices can be numbered the thyratron, the ignatron, the voltage regulator tube, various noise sources, and the transmit-receive radar tube. Fortunately, it was not necessary to have a detailed knowledge of the plasma state to be able to work with these devices.

Noise in the form of oscillations and instabilities is present in almost all plasma devices. In the past this noise was a villain to be exiled. Now, with a knowledge of the plasma state, devices can be built which amplify and modulate high-frequency waves by using those processes which previously caused the undesirable noise.

A plasma—under the proper biasing conditions—is, in itself, a slow-wave structure for electromagnetic waves. In a plasma traveling-wave tube, the velocity of the electromagnetic radiation is slowed to the speed at which the waves can most efficiently interact with an electron beam. The result of this interaction is the amplification of the electromagnetic wave. An advantage of this scheme over other slow-wave structures is that the properties of the amplifier can be changed by varying the properties of the plasma.

The interaction of a modulated electron beam with a plasma, under proper conditions, results in the growth of the modulation. The nonlinearities in plasmas can be used to create plasma parametric amplifiers. Coherent plasma oscillations can produce very short wavelength (millimeter) electromagnetic radiation. Plasmas can be operated as lasers. These are but a few of the uses of the plasma state as a control and an amplifier, that is, as a plasma tube.

4. For a subject which has been out of style for quite a few years, electric power generation has received a significant amount of attention from plasma researchers. The ultimate goal of the controlled-thermonuclear-reaction research is, of course, the production of energy. The most desirable end product of this energy-production process is electricity. It would be sad indeed if the steam cycle (thermonuclear energy in the form of heat, vaporizing water, driving steam turbines, generating voltages and currents) were necessary in order to light our plasma lamps. The direct conversion of thermonuclear energy into electrical energy is a subject of great interest. Since the motions of charged particles are electrical current, it appears possible to extract electrically by induction some of the energy of these currents.

The plasma-filled thermionic diode is a device which can convert heat energy into electrical energy. If the work functions of the anode and cathode are properly adjusted, an electron thermionically emitted from the cathode of the diode may traverse the plasma and arrive at the anode without an external voltage being applied across the diode. A current will flow through a load without a voltage source in the circuit.

If a plasma is forced to move perpendicular to a d-c magnetic field, an electric field is generated within the plasma in a direction normal to both the magnetic field and the plasma motion. The electric field generated in the plasma can be used to cause electrical current to flow in an external circuit as well as within the plasma. Energy is transferred to the load at the expense of the kinetic energy of the plasma particles. The advantage of the d-c magnetohydrodynamic generator is that it requires only moving plasma, not moving parts. Among the difficulties of such a scheme of power generation are the generation and maintenance of a plasma with appropriate characteristics.

5. Although the tail end of a conventional missile spews forth a very hot plasma, this is not what is meant by "plasma propulsion." Plasma propulsion is the name attached to various *low*-thrust techniques, techniques which produce thrusts per weight several orders of magnitude lower than chemical rockets. Who needs such a thing, one may ask. In the reaches of space, outside of the grasp of strong gravitational forces, there is little need for a high-thrust device. It may take a while, but the slow, steady, directional emission of ions and electrons from a rocket will accelerate that rocket to satisfactory speeds.

An ionic rocket engine, in its simplest form, consists of an electrostatic accelerator which ejects the ions from the rocket. The ejection of ions is the mass flow which drives the spaceship forward (action equals reaction). Of course it is not all that simple, for what is to prevent the ions, once ejected, from turning around and making for home, that is, back to the negative charges that they have left behind?

They will, in fact. After ejection, the positive ions must be combined with electrons or negative ions so that the mass which flows has charge neutrality. Briefly, for an ion engine to work, the "exhaust" must be charge-neutral.

To eliminate in part the space-charge problems of the ion engine, and to create numerous new problems, one can devise combinations of electric and magnetic fields which will eject from the rocket a neutral collection of charged particles. In general, neutral plasma or magnetohydrodynamic propulsion is capable of creating more thrust per weight than ionic propulsion. There are numerous schemes for the exhaust of a plasma at high speed. In all the schemes the force of ejection on the positive particles must be in the same direction as the force on the negative particles. This restriction limits severely the type of fields which can be used. Various forms of perpendicular electric and magnetic fields are perhaps the most popular field configurations.

Now let us get to the business of this book—a mathematical and physical look at the fundamental concepts of plasma physics.

BIBLIOGRAPHY

Bachynski, M. P.: Plasma Physics: An Elementary Review, *Proc. IRE*, **49**:1751 (1961).

Bishop, A. S.: "Project Sherwood: The U.S. Program in Controlled Fusion," Addison-Wesley Publishing Company, Inc., Reading, Mass., 1958.

Brown, S. C.: Plasma Physics, *Am. Scientist*, **50**:59 (1962).

Cambel, A. B.: Magneto-gasdynamics: Its Science and Technology, *Am. Scientist*, **50**:375 (1962).

Hellund, E. J.: "The Plasma State," Reinhold Publishing Corporation, New York, 1961.

Post, R. F.: Controlled Fusion Research: An Application of the Physics of High Temperature Plasmas, *Rev. Mod. Phys.*, **28**:338 (1956).

———: High Temperature Plasma Research and Controlled Fusion, *Ann. Rev. Nucl. Sci.*, **9**:367 (1959).

Plasma Physics Symposium held by the New York State Section of the American Physical Society on Apr. 13 and 14, 1962, published in:
Phys. Today, **15**(12): 21–54 (1962); **16**(3): 24–37 (1963); **16**(4): 26–54 (1963).

chapter

2

Phase Space, Distribution Functions, and Average Values

2-1 Introduction

A plasma is composed of a relatively large number of charged and uncharged particles. It is impractical to describe mathematically the behavior of each individual particle as it interacts with all other particles. Instead, it is sufficient to adopt a statistical approach: to ask what behavior is most likely for the particles. In this chapter we shall discuss statistical properties of systems containing large numbers of particles. We shall concentrate our attention on one of the most important (from a plasma point of view) systems that can be analyzed statistically, a system in which the particles have only translational kinetic energy and have potential energy only as a function of position coordinates. The system could be a collection of electrons, a monatomic gas, the oxygen ions in a lightning plasma, or a large number of colliding marbles in a gravitational field. In any event, the system must contain a large number of identical particles.

2-2 Phase Space

A complete specification of the state of the system, from a classical viewpoint, necessitates a knowledge of the position and velocity of each of the particles which comprise the system. The six quantities x, y, z, u_x, u_y, u_z must be specified for each particle. Let us invent a six-dimensional or "phase" space whose coordinates are x, y, z, u_x, u_y, u_z. We are at liberty to invent whatever we wish, as long as we do not contradict ourselves. Each particle in the system we are considering can be represented by a point in this six-dimensional space. The coordinates of the representative point give the velocity and position of the particle. There are exactly as many representative points in the six-dimensional space as there are particles in the system.

We shall consider now a differential volume in phase space

$$d\mathcal{V} = dx\,dy\,dz\,du_x\,du_y\,du_z$$

Within this small volume there are a large number of representative points. If this is not true, the system is not amenable to statistical analysis. The number of representative points within the differential volume at time t we shall call $d^6N(x,y,z,u_x,u_y,u_z,t)$. The number of representative points d^6N depends on time and on the location of the small volume in phase space. For example, a differential volume whose velocity coordinates are very large could be expected to contain relatively few representative points since there will be relatively few particles in any system with very large velocities. The density of representative points in phase space is given by

$$f(x,y,z,u_x,u_y,u_z,t) = \frac{d^6N(x,y,z,u_x,u_y,u_z,t)}{dx\,dy\,dz\,du_x\,du_y\,du_z} \tag{2-1}$$

where f will be called the "phase-space distribution function." Alternatively, Eq. (2-1) can be written

$$d^6N(x,y,z,u_x,u_y,u_z,t) = f(x,y,z,u_x,u_y,u_z,t)\,dx\,dy\,dz\,du_x\,du_y\,du_z \tag{2-2}$$

where d^6N gives the number of particles in the system at time t whose position coordinates are between x and $x + dx$, y and $y + dy$, z and $z + dz$ and whose velocities are between u_x and $u_x + du_x$, u_y and $u_y + du_y$, u_z and $u_z + du_z$.

One of the primary problems of statistical mechanics is to determine the form of the phase-space distribution function. The knowledge of this function yields, in a statistical sense, a complete description of the system.

It should be emphasized that in adopting a statistical approach we are not asking what each individual particle is doing. Rather, we are asking how many particles are in each differential position and velocity range at a given time.

2-3 Distribution Functions

There are basically two ways to obtain the phase-space distribution function for a system of particles: (1) by solving the Boltzmann equation, an equation which describes the behavior of the distribution function (we shall derive the Boltzmann equation in Chap. 4), and (2) by using statistical techniques based on the principle of maximization of the entropy or random character of the system. The latter approach is valuable only under equilibrium conditions, that is, when the system characteristics do not depend on time and when in any differential position space volume as many particles are moving in any one direction with a certain speed as are moving in the opposite direction with the same speed.

We shall not derive any distribution functions in this text. We shall write down without proof a few of the more common classical distribution functions and discuss their uses. We use the word "classical" to distinguish between the systems we will discuss and those which exhibit quantum-mechanical characteristics. The reader is referred to the bibliography for the detailed derivations of the distribution functions.

The most common classical distribution function is the Maxwell-Boltzmann distribution function. Maxwell-Boltzmann statistics describes the equilibrium state. It can be shown by either of the two techniques discussed above that, if the motion of the particles of any classical system is perfectly random, that system can be described by a Maxwell-Boltzmann distribution function. The statistical results say nothing about the form of the particle interactions. The effect of the particle interactions is to produce the random character of the system. It therefore is reasonable that the type of interaction not affect the resultant equilibrium statistics. The Maxwell-Boltzmann distribution function for a system of particles with translational kinetic energy and potential energy only as a function of position coordinates is

$$f(x,y,z,u_x,u_y,u_z) = \frac{N \exp\left[-\dfrac{m}{2\kappa T}(u_x^2+u_y^2+u_z^2) - \dfrac{\Phi(x,y,z)}{\kappa T}\right]}{\iiiint\!\!\iint \exp\left[-\dfrac{m}{2\kappa T}(u_x^2+u_y^2+u_z^2) - \dfrac{\Phi(x,y,z)}{\kappa T}\right] dx\,dy\,dz\,du_x\,du_y\,du_z} \tag{2-3}$$

where N = total number of particles in the system
m = mass of each of the particles of the system
T = absolute temperature of the system
κ = Boltzmann constant
$\Phi(x,y,z)$ = potential energy of the system as a function of position.

The integrals cover the total physical volume of the system and all velocities from $-\infty$ to $+\infty$.

The Maxwell-Boltzmann statistics is strictly valid only for equilibrium systems. The distribution function of Eq. (2-3) is, however, widely used in nonequilibrium situations. There are two reasons for this: (1) The system may be only slightly disturbed from equilibrium. Under these conditions, Maxwell-Boltzmann statistics may be a valid first approximation. (2) The actual distribution function of the system may be very difficult to determine. If some statistics must be assumed, the Maxwell-Boltzmann is a likely choice because it can be mathematically manipulated with relative ease.

If the system we are considering is composed of N electrons of mass m_e and charge $-|e|$ which interact with a gas containing n_g atoms per cubic meter of mass m_g in the presence of a d-c electric field $\mathbf{E}$, the system of electrons can be described by the Druyvesteyn distribution function

$$f(x,y,z,u_x,u_y,u_z) = \frac{N \exp\left[-\dfrac{3m_e}{m_g}\left(\dfrac{n_g\sigma}{|e|E}\right)^2 \mathcal{E}^2\right]}{\displaystyle\iiiint\!\!\iint_{\substack{\text{all } \mathbf{u} \\ \text{vol of system}}} \exp\left[-\dfrac{3m_e}{m_g}\left(\dfrac{n_g\sigma}{|e|E}\right)^2 \mathcal{E}^2\right] dx\,dy\,dz\,du_x\,du_y\,du_z} \tag{2-4}$$

where the assumption is made that the interacting particles behave like hard spheres. The collision cross section σ which appears in Eq. (2-4) is a measure of the probability of an electron–gas-atom collision. The collision cross section will be defined and discussed in Chap. 10. The Druyvesteyn distribution function falls off more rapidly at high electron kinetic energies, $\mathcal{E} = \frac{1}{2}m(u_x^2 + u_y^2 + u_z^2)$, than does the Maxwell-Boltzmann. Thus there are fewer high-energy particles in a system which can be described by a Druyvesteyn distribution function than there are in a comparable Maxwell-Boltzmann system.

If the system of electrons is under the influence of an alternating electric field of angular frequency ω, the system can be described by the following distribution function:

$$f(x,y,z,u_x,u_y,u_z) = \frac{N \exp\left\{-\dfrac{3m_e}{m_g}\left(\dfrac{n_g\sigma}{|e|E_{\text{rms}}}\right)^2\left[\mathcal{E}^2 + \left(\dfrac{\omega}{n_g\sigma}\right)^2 m_e\mathcal{E}\right]\right\}}{\displaystyle\iiiint\!\!\iint_{\substack{\text{all } \mathbf{u} \\ \text{vol of system}}} \exp\left\{-\dfrac{3m_e}{m_g}\left(\dfrac{n_g\sigma}{|e|E_{\text{rms}}}\right)^2\left[\mathcal{E}^2 + \left(\dfrac{\omega}{n_g\sigma}\right)^2 m_e\mathcal{E}\right]\right\} dx\,dy\,dz\,du_x\,du_y\,du_z} \tag{2-5}$$

where E_{rms} is the rms value of the electric field. At low frequencies

Eq. (2-5) yields the Druyvesteyn distribution function. At high frequencies the second term in the exponent is dominant, and Eq. (2-5) has the energy dependence of a Maxwell-Boltzmann distribution function.

The distribution functions of Eqs. (2-4) and (2-5) can be derived from the Boltzmann equation. It is necessary to make a number of approximations to effect a solution. The solutions are not valid if the approximations are not valid. The two primary assumptions made are that (1) the interacting particles behave like hard spheres and (2) the energy of the electrons is much greater than the thermal energy of the gas. Further, Eqs. (2-4) and (2-5) are approximations in that each should contain a small additive correction term. Physically, the small correction term must be anisotropic to account for the small directional motion of the electrons due to the applied electric field. It is the additive terms which yield the electron current density. The correction terms are derived in the references.

Whichever statistics is valid, the distribution function expresses the density of representative points in phase space. It is sometimes valuable to obtain a distribution function which depends on several, but not all, the coordinates of the six-dimensional phase space. For instance, if we wish to know the number of particles in a system which, irrespective of u_z, has a velocity in the x direction between u_x and $u_x + du_x$, a velocity in the y direction between u_y and $u_y + du_y$, and position coordinates between x and $x + dx$, y and $y + dy$, z and $z + dz$, we must add up the representative points in all small volumes of phase space whose x, y, z, u_x, u_y coordinates are constant but whose u_z coordinate varies between $-\infty$ and $+\infty$. The addition is done by integration.

$$d^5N = \left[\int_{-\infty}^{+\infty} f(x,y,z,u_x,u_y,u_z,t)\, du_z\right] dx\, dy\, dz\, du_x\, du_y \qquad (2\text{-}6)$$

If we wish to know the number of particles of the system in a small volume in real space, we must count up all of the representative points in phase space whose x, y, z coordinates are fixed but whose u_x, u_y, u_z coordinates vary from $-\infty$ to $+\infty$.

$$d^3N = \left[\iiint\limits_{\text{all } \mathbf{u}} f(x,y,z,u_x,u_y,u_z,t)\, du_x\, du_y\, du_z\right] dx\, dy\, dz \qquad (2\text{-}7)$$

The density of particles in real space is therefore given by

$$n(x,y,z,t) = \iiint\limits_{\text{all } \mathbf{u}} f(\mathbf{r},\mathbf{u},t)\, d^3u \qquad (2\text{-}8)$$

where we adopt the following shorthand notation: d^3u represents a differential "volume," each of whose three sides has the dimension of velocity; d^3r represents a differential volume in real space; and $(\mathbf{r},\mathbf{u},t)$ signifies the dependence of the distribution function on three velocity coordinates, three position coordinates, and time.

As an example of the application of Eq. (2-8), we shall calculate the density of particles in real space assuming a Maxwell-Boltzmann distribution. The substitution of Eq. (2-3) into Eq. (2-8) yields

$$n(x,y,z) = \frac{N \exp\left[-\Phi(x,y,z)/\kappa T\right]}{\displaystyle\iiint_{\text{vol of system}} \exp\left[-\Phi(x,y,z)/\kappa T\right]\,dx\,dy\,dz} \tag{2-9}$$

The factor

$$\frac{N}{\displaystyle\iiint_{\text{vol of system}} \exp\left[-\Phi(x,y,z)/\kappa T\right]\,dx\,dy\,dz}$$

is a constant which can be evaluated most easily by substituting $\Phi = 0$ in the numerator of Eq. (2-9). The constant is thus evaluated as the particle density at positions where the potential energy of the particles is zero.

$$n(x,y,z) = n_{\Phi=0} \exp\left[-\frac{\Phi(x,y,z)}{\kappa T}\right] \tag{2-10}$$

We shall use Eq. (2-10) in later chapters.

It is often useful to know the number of particles in a system whose *speed*, $u = (u_x^2 + u_y^2 + u_z^2)^{1/2}$, is between u and $u + du$ and whose position coordinates are between x and $x + dx$, y and $y + dy$, z and $z + dz$. To find this number of particles we must add up all the representative points in the small volumes of phase space whose position coordinates are constant and whose velocity coordinates yield the required speed $u = (u_x^2 + u_y^2 + u_z^2)^{1/2}$.

$$d^4N = \iint_{\mathbf{u}}^{u=(u_x{}^2+u_y{}^2+u_z{}^2)^{1/2}} f(\mathbf{r},\mathbf{u},t)\,d^3u\,d^3r \tag{2-11}$$

Equation (2-11) can be easily evaluated by expressing the velocity variables in *spherical* velocity coordinates.

$$f(\mathbf{r},\mathbf{u},t) \rightarrow f(\mathbf{r},u,\theta,\phi,t)$$
$$d^3u \rightarrow u^2 \sin\theta\,d\theta\,d\phi\,du$$

Equation (2-11) can then be expressed as

$$d^4N = \left[\int_{\theta=0}^{\pi} \int_{\phi=0}^{2\pi} f(\mathbf{r},u,\theta,\phi,t) \sin\theta \, d\theta \, d\phi \right] u^2 \, du \, d^3r \qquad (2\text{-}12)$$

where speed is held constant and all velocities which have the same speed are included by integration over the angular velocity coordinates. One special case is worthy of note. If the distribution function for the particles is isotropic, that is, if the distribution function is independent of θ and ϕ, the integrations in Eq. (2-12) can be performed with the result that

$$d^4N = 4\pi u^2 f(\mathbf{r},u,t) \, du \, d^3r \qquad (2\text{-}13)$$

Equation (2-13) can be rewritten

$$d^4N = h(\mathbf{r},u,t) \, du \, d^3r \qquad (2\text{-}14)$$

where $h(\mathbf{r},u,t) = 4\pi u^2 f(\mathbf{r},u,t)$. The quantity $h(\mathbf{r},u,t)$ will be called the "isotropic speed distribution function." It represents a density of points in a four-dimensional space whose coordinates are u and three real-space coordinates. We shall use the isotropic speed distribution function in Chap. 3.

2-4 Average Values

If five people take a certain test and if two people receive a grade of 100, two people a grade of 90, and the remaining person a grade of 70, the average grade of the class is calculated as follows: We multiply each grade received by the number of people receiving that particular grade, add the weighted grades, and divide by the total number of people receiving grades. The average grade for the class of five is therefore

$$\frac{2 \times 100 + 2 \times 90 + 1 \times 70}{5} = 90$$

In a similar fashion we can calculate the average values of the parameters of a plasma. We shall calculate the average kinetic energy of one group of identical particles (either electrons, one type of ions, or one type of neutrals) in a plasma. The kinetic energy of the particles is $\mathcal{E} = \frac{1}{2}m(u_x^2 + u_y^2 + u_z^2)$. Each differential volume in phase space has a particular energy associated with it; that is, all points within a small volume of phase space represent particles which have approximately the same kinetic energy, the kinetic energy being determined by the u_x, u_y, u_z coordinates of the differential volume. The number of representative points within a small volume whose phase-space coordinates are u_x, u_y, u_z, x, y, z is given by Eq. (2-2) if f is the appropriate particle distribution function. To find the average kinetic energy of the particles we must multiply each possible energy by the number of particles carrying that energy, add the weighted energies, and divide by the total number of plasma particles we are considering. The addition of the weighted averages is accomplished by integration over all of phase space. The average kinetic energy of the particles therefore can be written

$$\overline{\mathcal{E}(t)} = \frac{\displaystyle\iiiint\!\!\iint_{\substack{\text{all } \mathbf{u} \\ \text{vol of system}}} [\tfrac{1}{2}m(u_x^2 + u_y^2 + u_z^2)][f(\mathbf{r},\mathbf{u},t)]\, d^3r\, d^3u}{\displaystyle\iiiint\!\!\iint_{\substack{\text{all } \mathbf{u} \\ \text{vol of system}}} [f(\mathbf{r},\mathbf{u},t)]\, d^3r\, d^3u} \tag{2-15}$$

The denominator of Eq. (2-15) is equal to the total number of plasma particles we are considering. For the case of Maxwell-Boltzmann statistics, Eq. (2-15) yields

$$\overline{\mathcal{E}} = \frac{\displaystyle\iiint_{\text{all } \mathbf{u}} [\tfrac{1}{2}m(u_x^2 + u_y^2 + u_z^2)] \left\{\exp\left[-\frac{m}{2\kappa T}(u_x^2 + u_y^2 + u_z^2)\right]\right\} du_x\, du_y\, du_z}{\displaystyle\iiint_{\text{all } \mathbf{u}} \exp\left[-\frac{m}{2\kappa T}(u_x^2 + u_y^2 + u_z^2)\right] du_x\, du_y\, du_z}$$

$$= \tfrac{3}{2}\kappa T \tag{2-16}$$

The average kinetic energy which we have written down in Eq. (2-15) is the average over all velocities and over all position coordinates. It is more often useful to know the average kinetic energy associated with those particles contained in a small position coordinate volume d^3r centered about $\mathbf{r}$. The average energy as a function of position is calculated in the same manner as before except that we do not integrate over the position coordinates.

$$\overline{\mathcal{E}(\mathbf{r},t)} = \frac{\displaystyle\iiint_{\text{all } \mathbf{u}} \tfrac{1}{2}m(u_x^2 + u_y^2 + u_z^2) f(\mathbf{u},\mathbf{r},t)\, d^3u}{\displaystyle\iiint_{\text{all } \mathbf{u}} f(\mathbf{u},\mathbf{r},t)\, d^3u} \tag{2-17}$$

Notice that for a system of particles described by Maxwell-Boltzmann statistics, $\overline{\mathcal{E}(\mathbf{r})}$ is independent of position.

The average velocity, which we shall call $\mathbf{v}$, of the particles contained in d^3r is given by

$$\mathbf{v}(\mathbf{r},t) \equiv \overline{\mathbf{u}(\mathbf{r},t)} = \frac{\iiint\limits_{\text{all } \mathbf{u}} \mathbf{u} f(\mathbf{u},\mathbf{r},t)\, d^3u}{\iiint\limits_{\text{all } \mathbf{u}} f(\mathbf{u},\mathbf{r},t)\, d^3u} \tag{2-18}$$

In an equilibrium plasma, the average velocity is zero since as many particles are going in any given direction with a certain velocity as are going opposite to that direction with the same magnitude of velocity.

The average speed of the plasma particles in d^3r is given by

$$\overline{u(\mathbf{r},t)} = \frac{\iiint\limits_{\text{all } \mathbf{u}} (u_x^2 + u_y^2 + u_z^2)^{1/2} f(\mathbf{u},\mathbf{r},t)\, d^3u}{\iiint\limits_{\text{all } \mathbf{u}} f(\mathbf{u},\mathbf{r},t)\, d^3u} \tag{2-19}$$

If the plasma particles have no preferred direction of motion, Eq. (2-19) can be transformed into spherical velocity coordinates and expressed in terms of the isotropic speed distribution function.

$$\overline{u(\mathbf{r},t)} = \frac{\int_0^\infty u h(\mathbf{r},u,t)\, du}{\int_0^\infty h(\mathbf{r},u,t)\, du} \tag{2-20}$$

In a similar manner, any average parameter of the particular plasma constituent under study can be obtained.

The "system" which we have been examining in the foregoing is not the whole plasma. The system is composed of one of the plasma constituents—electrons, one type of ions, or one type of neutral particles. The interaction of this system of particles with all the other particles in the plasma determines the phase-space distribution function for the system. Once this distribution function has been obtained, we need no longer concern ourselves with the behavior of the other member constituents of the plasma. In other words, the plasma can be considered to be composed of several separate systems, each with its own particular distribution function.

PROBLEMS

2-1 Calculate the average velocity, average speed, and average energy of a system of particles which obeys Maxwell-Boltzmann statistics if (*a*) $\Phi(x,y,z) = 0$ and (*b*) $\Phi(x,y,z) = mgz$.

2-2 Determine numerical values for the average velocity, average speed, and average energy found in Prob. 2-1*a* if (*a*) the particles are electrons and (*b*) the particles are oxygen ions (O^+) for $T = 0°K$, $T = 300°K$ (86°F), and $T = 1000°K$. Express the values of average energy both in joules and in electron volts (1 ev = 1 volt times 1.6×10^{-19} coul; the energy an electron gains in falling through a potential difference of 1 volt).

2-3 Consider a four-dimensional space whose coordinates are $\mathcal{E}$, x, y, z. Define an energy distribution function $g(\mathcal{E},x,y,z)$ for the four-dimensional space. Express the energy distribution function in terms of the appropriate phase-space distribution function $f(u_x,u_y,u_z,x,y,z)$. Consider only the case of an isotropic phase-space distribution function.
HINT: $d\mathcal{E} = mu\, du$.

2-4 Plot the phase-space distribution function, the speed distribution function, and the energy distribution function for a system of particles which obeys Maxwell-Boltzmann statistics. Compare the most probable values of velocity, speed, and energy with the average values calculated in Prob. 2-1*a*. Discuss.

2-5 Using the energy distribution function found in Prob. 2-3, calculate the average energy of a system of particles which obeys Maxwell-Boltzmann statistics. Let $\Phi(x,y,z) = 0$.

2-6 Find the number of particles in a Maxwell-Boltzmann system of N particles which have speeds greater than the average speed. Compute the average speed of those particles which have speeds greater than the average speed of the total system.

2-7 Assume that the speed distribution function for a group of N particles contained in a volume V is

$$h(u,x,y,z) = ku^2 \qquad a > u > 0$$
$$h(u,x,y,z) = 0 \qquad u > a$$

(*a*) Plot $h(u,x,y,z)$ as a function of u.
(*b*) Find the constant k in terms of N and a.
(*c*) Compute the average energy and the most probable energy of the particles.

BIBLIOGRAPHY

Chapman, S., and T. G. Cowling: "The Mathematical Theory of Non-uniform Gases," Cambridge University Press, New York, 1958.

p. 69—On the derivation of Maxwell-Boltzmann statistics from the Boltzmann equation.

Druyvesteyn, M. J.: Bemerkungen zu zwei früheren Arbeiten über die Electronendiffusion, *Physica,* **1**:1003 (1934).
On the original derivation of the Druyvesteyn distribution function.

Kennard, E. H.: "Kinetic Theory of Gases," McGraw-Hill Book Company, Inc., New York, 1938.

Margenau, H.: Conduction and Dispersion of Ionized Gases at High Frequencies, *Phys. Rev.*, **69**:508 (1946).
On the derivation of Eq. (2-5) with correction term from the Boltzmann equation.

Morse, P. M., W. P. Allis, and E. S. Lamar: Velocity Distributions for Elastically Colliding Electrons, *Phys. Rev.*, **48**:412 (1935).
On the derivation of the Druyvesteyn distribution function with correction term from the Boltzmann equation.

Present, R. D.: "Kinetic Theory of Gases," McGraw-Hill Book Company, Inc., New York, 1958.

Sears, F. W.: "An Introduction to Thermodynamics, the Kinetic Theory of Gases, and Statistical Mechanics," Addison-Wesley Publishing Company, Inc., Reading, Mass., 1953.
p. 276—On the derivation of Maxwell-Boltzmann statistics using the maximization-of-entropy principle.

chapter

3

Random Current Density and Kinetic Pressure

3-1 Introduction

The random current density of a system of particles is defined as the rate at which particles cross a unit area from one side only. If the system of particles is described by an isotropic distribution function, the random current density will be independent of the orientation of the unit area. The total particle current density within the system of particles is therefore equal to zero since as many particles will cross a given unit area from one side as will cross from the opposite side. The concept of a random current density is useful in discussing the interaction of a plasma with a material body, that is, in determining the properties of the plasma sheath, as we shall see in Chap. 9. The random current density is a current of particles only. It is not an electrical current unless the particles carry a charge. In this case the random electrical current density is found by multiplying the random current density by the charge which the particles carry.

The kinetic pressure of a system of particles is a measure of the force exerted by the random motion of the particles. The origin of both random current density and kinetic pressure is the random motion of the particles. We should expect the kinetic pressure of a system of particles to be proportional to the mass of the particles since the force exerted by the particles must be proportional to their mass. We should further expect on physical grounds that both the random current density and the kinetic pressure would be proportional to the density of the particles and to some function of the speed of the particles. The kinetic pressure of a system of particles plays an important role in determining the behavior of the particles under external stimuli. We shall find that in order to contain a plasma it is necessary to oppose the pressure inherent in the plasma with electromagnetic forces.

3-2 Random Current Density

In our discussion of random current density, we shall use the particle distribution function expressed in spherical velocity coordinates $f(\mathbf{r},u,\theta,\phi,t)$. We shall make the assumptions that (1) the distribution function is independent of position coordinates and (2) the distribution function describes an equilibrium isotropic system. The particle phase-space distribution function can therefore be written $f(u)$.

If we place a small area dA within the system of particles, the area will be bombarded in time dt by groups of particles having many different velocities. We shall label one of the bombarding groups of particles "$u\theta\phi$ particles." These particles have a speed between u and $u + du$ and are moving in a direction defined by the spherical coordinate angles between θ and $\theta + d\theta$, ϕ and $\phi + d\phi$. We wish to calculate the number of $u\theta\phi$ particles which pass through dA in time dt. We shall then add up all particles whose speeds and directions of motion are such that the particles will pass through dA from one side in time dt. The addition will be accomplished by integration.

The number of $u\theta\phi$ particles which pass through dA in time dt is equal to the number of $u\theta\phi$ particles contained in the volume of ordinary space with base dA and slant height $u\,dt$ shown in Fig. 3-1. A $u\theta\phi$ particle at the top of the volume will just reach dA in time dt. All $u\theta\phi$ particles within the volume will pass through dA in time dt. There are other groups of particles within the volume which will also pass through dA, but these will not concern us since we shall take them into account when we add the contribution of all groups of particles to the random current density. To reiterate, the number of $u\theta\phi$ particles which will pass through dA in time dt is the number of $u\theta\phi$ particles contained in the volume $dA\,u\,dt\cos\theta$ in ordinary space.

The number of $u\theta\phi$ particles in a volume $dx\,dy\,dz$ of ordinary space is given by

$$d^6N_{u\theta\phi} = f(u)u^2 \sin\theta\, d\theta\, d\phi\, du\, dx\, dy\, dz \qquad (3\text{-}1)$$

Since we have assumed that the number of $u\theta\phi$ particles is the same at any position in xyz coordinates, the number of $u\theta\phi$ particles in the

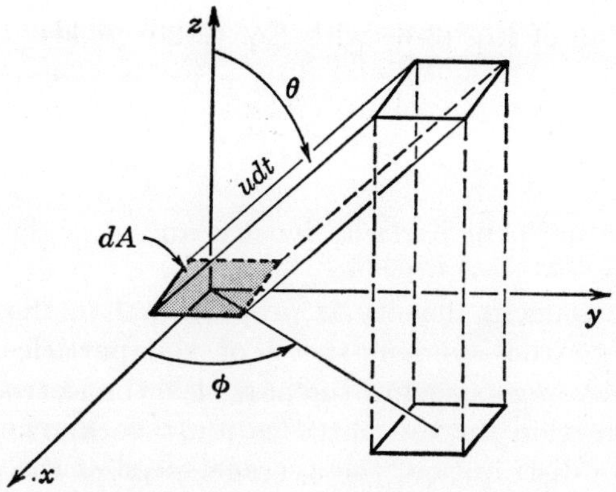

Fig. 3-1 Volume containing all $u\theta\phi$ particles which pass through dA in time dt. The small area dA lies in the xy plane.

real-space volume $dA\ u\ dt \cos\theta$ is given by

$$dT_{u\theta\phi} = f(u)u^2 \sin\theta\ d\theta\ d\phi\ du\ dA\ u\ dt \cos\theta \tag{3-2}$$

The random current density due to $u\theta\phi$ particles is found by dividing the number of $u\theta\phi$ particles which hit dA in time dt, $dT_{u\theta\phi}$, by $dA\ dt$.

$$dJ_{u\theta\phi} = f(u)u^3 \sin\theta \cos\theta\ d\theta\ d\phi\ du \tag{3-3}$$

It is convenient to substitute for the isotropic phase-space distribution function the isotropic speed distribution function $h(u) = 4\pi u^2 f(u)$.

$$dJ_{u\theta\phi} = \frac{uh(u)}{4\pi} \sin\theta \cos\theta\ d\theta\ d\phi\ du \tag{3-4}$$

The total random current density is found by adding up all particles whose directions of motion vary between $0 \le \theta \le \pi/2$ (particles approaching dA from one side only) and $0 \le \phi \le 2\pi$ and whose speeds vary between zero and infinity.

$$J = \int_{u=0}^{\infty} \int_{\phi=0}^{2\pi} \int_{\theta=0}^{\pi/2} \frac{uh(u)}{4\pi} \sin\theta \cos\theta\ d\theta\ d\phi\ du \tag{3-5}$$

The result of the integration over the angles is

$$J = \frac{\int_0^{\infty} uh(u)\ du}{4} \tag{3-6}$$

The combination of Eq. (3-6) with Eq. (2-20) yields

$$J = \frac{n\bar{u}}{4} \tag{3-7}$$

where n is the uniform particle density and $\bar{u}$ is the average speed of the particles.

The random current density is proportional to the density of the particles and to the average speed of the particles. In order to determine the average speed of the particles, the isotropic phase-space distribution function for the particles must be known. For a Maxwell-Boltzmann distribution, the average speed of the particles is

$$\bar{u} = \left(\frac{8\kappa T}{\pi m}\right)^{1/2} \tag{3-8}$$

In this case the random current density is directly proportional to the square root of the temperature of the particles and inversely proportional to the square root of the mass of the particles.

3-3 Kinetic Pressure

Kinetic pressure is the force per unit area that a system of particles exerts because of the random motion of the particles. We shall make the same assumptions about the form of the distribution function that we made in discussing the random current density. We shall consider the area dA of Fig. 3-1 to be solid and shall calculate the force on dA due to particle bombardment. This force, when divided by dA, will yield the kinetic pressure of the bombarding particles.

The number of $u\theta\phi$ particles which hit dA in time dt is given by Eq. (3-2). Each of these particles carries a momentum equal in magnitude to mu. We shall assume that a $u\theta\phi$ particle hitting dA behaves like a marble striking a flat piece of sidewalk. Figure 3-2 shows the result of a collision of a $u\theta\phi$ particle with dA. The angle of reflection is equal to the incident angle θ of the $u\theta\phi$ particle. The momentum component of the incoming particle perpendicular to dA is $mu \cos\theta$. Upon reflection, the momentum component of the $u\theta\phi$ particle is still $mu \cos\theta$, but it is in the opposite direction. Since momentum must be conserved, the area dA will have received a momentum $2mu \cos\theta$ during the collision.

The total momentum transfer to the small area dA in time dt by $u\theta\phi$ particles is found by multiplying the momentum transfer from

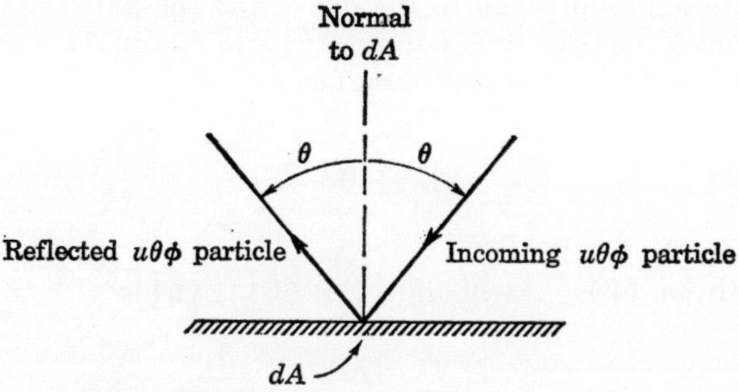

Fig. 3-2 Trajectory of a $u\theta\phi$ particle striking dA.

one $u\theta\phi$ particle by the number of $u\theta\phi$ particles which hit dA in dt.

$$dS_{u\theta\phi} = \left[\frac{uh(u)}{4\pi} \sin\theta \cos\theta \, d\theta \, d\phi \, du \, dA \, dt\right] (2mu \cos\theta) \tag{3-9}$$

Force is the momentum change per unit time so that $dF_{u\theta\phi} = dS_{u\theta\phi}/dt$. The pressure due to the $u\theta\phi$ particles is $dF_{u\theta\phi}/dA$.

$$dP_{u\theta\phi} = \frac{mu^2h(u)}{2\pi} \cos^2\theta \sin\theta \, d\theta \, d\phi \, du \tag{3-10}$$

The total pressure is found by integration, where again the angle θ varies between 0 and $\pi/2$.

$$P = \int_{u=0}^{\infty} \int_{\phi=0}^{2\pi} \int_{\theta=0}^{\pi/2} \frac{mu^2h(u)}{2\pi} \cos^2\theta \sin\theta \, d\theta \, d\phi \, du \tag{3-11}$$

After performing the integration over the angles, we find that

$$P = \frac{m}{3} \int_0^{\infty} h(u)u^2 \, du \tag{3-12}$$

The integral in Eq. (3-12) yields the average squared speed times the uniform density of the particles

$$P = \frac{nm}{3} \overline{u^2} \tag{3-13}$$

The pressure is proportional to the density of the particles, the mass of the particles, and the average squared speed. If the particles are distributed in a Maxwell-Boltzmann fashion,

$$\overline{u^2} = \frac{3\kappa T}{m} \tag{3-14}$$

The substitution of Eq. (3-14) into Eq. (3-13) yields

$$P = n\kappa T \tag{3-15}$$

Equation (3-15) expresses the relationship between pressure, density, and temperature for a Maxwell-Boltzmann system of particles. Equation (3-15) is known as the "ideal gas law."

PROBLEMS

3-1 Find the number of molecules per unit volume in a gas at 300°K if the pressure is (*a*) 1 mm Hg, (*b*) 10^{-3} mm Hg, and (*c*) 10^{-6} mm Hg.
NOTE: The number of molecules in a kilogram-mole is 6.02×10^{26}. Under standard conditions (760 mm Hg, 273°K), 1 kg-mole occupies 22.4 m³.

3-2 Determine numerical values for the random current density of a group of electrons which obeys Maxwell-Boltzmann statistics if
(*a*) P = 760 mm Hg T = 0°K, 300°K, 1000°K
(*b*) P = 1 mm Hg T = 0°K, 300°K, 1000°K
(*c*) P = 10^{-3} mm Hg T = 0°K, 300°K, 1000°K
(*d*) P = 10^{-6} mm Hg T = 0°K, 300°K, 1000°K
Repeat the above for a Maxwell-Boltzmann gas of oxygen ions.

3-3 Derive expressions for the pressure and for the random current density in a two-dimensional spatially uniform isotropic gas. In a two-dimensional gas the gas particles are constrained to move only in a plane.

3-4 A closed flask of volume V containing a Maxwell-Boltzmann gas is placed in a vacuum chamber. A very small opening dA is made in the flask so that gas may escape. Assume that the escape of gas is sufficiently slow so that the Maxwell-Boltzmann distribution of gas particles within the flask is not disturbed. Find the number of particles with speeds between u and $u + du$ that leave the flask per unit time. Calculate the average speed of the particles leaving the flask. Compare this average speed with the average speed of the particles within the flask.

3-5 Find the time required for the pressure within the flask of Prob. 3-4 to decrease to one-third its original value.

BIBLIOGRAPHY

Kennard, E. H.: "Kinetic Theory of Gases," McGraw-Hill Book Company, Inc., New York, 1938.

Sears, F. W.: "An Introduction to Thermodynamics, the Kinetic Theory of Gases, and Statistical Mechanics," Addison-Wesley Publishing Company, Inc., Reading, Mass., 1953.
chap. 11—On the derivation of the expressions for the random current density and the kinetic pressure.

chapter

4

The Equations of Plasma Physics

4-1 Introduction

One of the objectives of physics and of modern engineering is to approximate nature by a series of mathematical relations. These equations are written so that they incorporate within themselves the aspects of phenomena that have been experimentally determined. Once the equations have been set down, they can be solved and used to predict the results of future measurements. The worth of a theory (or mathematical model) is its ability to predict the results of future experiments. If the predictions of the mathematical equations are not justified by experiment, a new mathematical model must be sought. All mathematical theories are, to some extent, approximations; nature has been unkind enough to see to that. It is not unreasonable, then, that different mathematical theories are devised to describe one set of physical observations. If the theories are different, the degree of approximation to nature will be different. This does not mean that the theory with the poorer degree of approximation is not a useful one. Newton's equation of motion is perfectly adequate for describing the behavior of a baseball. It is not, however, so accurate as the quantum-mechanical theory of baseballs. Often, accurate theories are not used because their accuracy is not warranted or their application is too difficult. In general, the more closely the mathematical theory approximates nature, the less susceptible the equations of that mathematical theory are to solution.

There are two widely used mathematical theories describing the plasma state (we exempt the quantum-mechanical theory from this book). The two theories involve (1) the use of the hydrodynamic equations, that is, the momentum transfer equation and the continuity equation, in conjunction with the equation of state and with Maxwell's equations and (2) the use of the Boltzmann equation in conjunction with Maxwell's equations. In this chapter the equations of these two mathematical approaches will be discussed. Later chapters of the book will be devoted to determining various solutions to these equations. The road to travel is clear; the solution of sets of partial differential and integrodifferential equations is not so simple. It will be necessary to make many assumptions and approximations along the way in order to arrive at answers. The validity of some assumptions will be justified only by the plausibility of the solution. At times we shall not even be this fortunate. Forewarned is forearmed.

Within a plasma there may be many types of positive and negative particles, as well as many types of neutral particles. The plasma created by a lightning stroke is a good example of a multicomponent plasma. In this chapter and throughout the remainder of the book we shall consider only the three-component plasma: a plasma composed of electrons, one type of positive ions, and one type of neutral particles. The extension of the mathematics of the three-component plasma to describe a multicomponent plasma is straightforward. Further, we shall consider only those cases in which there is no production or annihilation of plasma electrons, plasma ions, or plasma neutrals.

4-2 Maxwell's Equations

Maxwell's equations describe mathematically the behavior of electric and magnetic fields and their sources: charges and currents. It is assumed that the reader is familiar with electromagnetic theory, so that, without further discussion, Maxwell's equations will be set down, both in differential and in integral form:

$$\nabla \cdot \mathbf{D} = \rho \qquad \oiint \mathbf{D} \cdot \mathbf{n}\, dA = Q \tag{4-1a,b}$$

$$\nabla \times \mathbf{H} = \mathbf{j} + \frac{\partial \mathbf{D}}{\partial t} \qquad \oint \mathbf{H} \cdot d\mathbf{L} = I + \iint \frac{\partial \mathbf{D}}{\partial t} \cdot \mathbf{n}\, dA \tag{4-2a,b}$$

$$\nabla \times \mathbf{E} = -\frac{\partial \mathbf{B}}{\partial t} \qquad \oint \mathbf{E} \cdot d\mathbf{L} = -\iint \frac{\partial \mathbf{B}}{\partial t} \cdot \mathbf{n}\, dA \tag{4-3a,b}$$

$$\nabla \cdot \mathbf{B} = 0 \qquad \oiint \mathbf{B} \cdot \mathbf{n}\, dA = 0 \tag{4-4a,b}$$

where $\mathbf{D} = \epsilon\mathbf{E}$ and $\mathbf{B} = \mu\mathbf{H}$.

The charge density ρ and the current density $\mathbf{j}$ can be determined in terms of the electric and magnetic fields from the "equations of motion" of the particles. Thus, whichever of the two modes of description for the motion of charged particles is used, the expressions for ρ and $\mathbf{j}$ as a function of electric and magnetic fields from that description can be substituted into Maxwell's equations. The resulting set of equations has as its unknowns only the electric and the magnetic fields.

It should be noted that in using ρ and $\mathbf{j}$ we shall be discussing the plasma from a macroscopic viewpoint. That is, we shall be ignoring the discreteness of the plasma particles and of the electric charges which those particles carry. We shall therefore be using Maxwell's equations to describe the macroscopic fields in the plasma and not to describe the microscopic interparticle fields.

4-3 Hydrodynamic Equations

4-3-1 Introduction and Assumptions. The physical basis for our derivation of the hydrodynamic equations is the assumption that a three-component plasma may be treated as if it were composed of three intermingled nonviscous *fluids:* The electrons comprise one fluid, the positive ions another, and the neutral particles the third.[1] The velocity of any one fluid at any point in space is assumed to be the *average* velocity characteristic of the particles comprising that fluid at that particular point. The three fluids interact with one another by virtue of electron-ion, electron-neutral, and ion-neutral collisions.

We shall derive a momentum transfer equation and a continuity equation for each of the three intermingled nonviscous fluids. We shall find that these six hydrodynamic equations contain nine unknown quantities. In order to be able to solve for the nine unknown quantities, it will be necessary to obtain three additional equations describing the plasma. These three additional equations will be specified in an arbitrary form, their exact formulation being a function of the thermodynamics of the particular plasma system under study.

4-3-2 The Momentum Transfer Equations. We shall consider first the derivation of the momentum transfer equation for the electron fluid. We shall determine the total force acting on the electron fluid within a differential volume $d^3r = dx\,dy\,dz$ shown in Fig. 4-1. This total force will then be equated to the mass of the electron fluid within the differential volume multiplied by its acceleration. The force due to the pressure of the electron fluid acting on the left-hand side of the volume is given by the pressure on the left-hand face multiplied by the area of the face $dF_{xl} = P_e(x,y,z)\,dy\,dz$. The force

[1] The mathematical justification for the fluid approach is given in Appendix A.

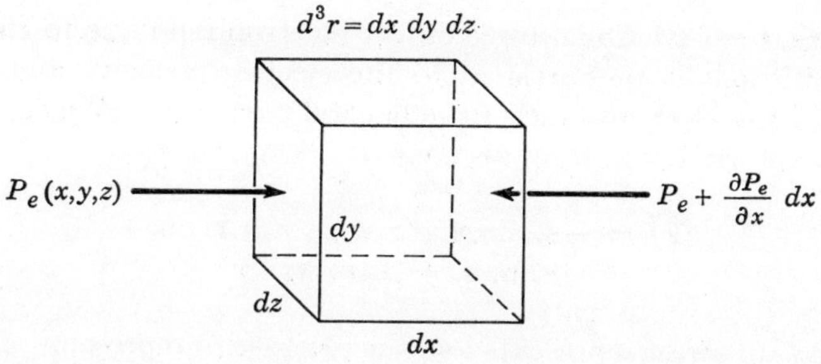

Fig. 4-1 A differential volume of electron fluid.

on the right-hand side of the volume due to pressure is given by $dF_{xr} = [P_e + (\partial P_e/\partial x)\,dx]\,dy\,dz$, where the electron pressure has been expanded in a Taylor series. The net pressure force in the x direction on the electron fluid in the differential volume is the difference between the force on the left and the force on the right, $dF_x = -(\partial P_e/\partial x)\,d^3r$. If similar consideration is taken of the pressure forces on the other two sets of faces of the differential volume, the total pressure force can then be written

$$d\mathbf{F}_p = -(\nabla P_e)\,d^3r \tag{4-5}$$

The force on one electron due to electric and magnetic fields is given by

$$\mathbf{F} = -|e|(\mathbf{E} + \mathbf{u}_e \times \mathbf{B}) \tag{4-6}$$

where $|e|$ is the magnitude of the electron charge, $\mathbf{u}_e$ is the electron velocity, and $\mathbf{E}$ and $\mathbf{B}$ are the electric field intensity and the magnetic flux density, respectively. If the density of electrons is $n_e(\mathbf{r},t)$, there are $n_e\,d^3r$ electrons in the volume d^3r. The total force on the electron fluid in d^3r due to electric and magnetic fields can then be expressed as

$$d\mathbf{F}_{EB} = -(|e|n_e\,d^3r)(\mathbf{E} + \mathbf{v}_e \times \mathbf{B}) \tag{4-7}$$

where $\mathbf{v}_e$ is the *average* velocity of the electrons in d^3r, that is, the velocity of the electron fluid.

The remaining force on the electrons is due to the interaction of the electron fluid with the ion fluid and with the neutral-particle fluid. We shall assume that the frictional or damping force on the electron fluid due to electron-ion collisions is proportional to the velocity of the electron fluid and to the densities of the electrons and of the ions and that the frictional force on the electron fluid due to electron-neutral collisions is proportional to the electron velocity and to the densities of the electrons and of the neutral particles.[1] We can therefore write the frictional-force term as

$$\begin{aligned} d\mathbf{F}_f &= -\alpha_{ei}n_i n_e \mathbf{v}_e \, d^3r - \alpha_{en}n_n n_e \mathbf{v}_e \, d^3r \\ &= -n_e(\alpha_{ei}n_i + \alpha_{en}n_n)\mathbf{v}_e \, d^3r \end{aligned} \tag{4-8}$$

where the i subscript represents ion, the e subscript represents electron, the n subscript represents neutral particle, and the α's are variable factors (assumed known) to take complete account of the collision interactions. In general, the factor α_{ei} may be a function of $\mathbf{v}_e$ and $\mathbf{v}_i$, and the factor α_{en} a function of $\mathbf{v}_e$ and $\mathbf{v}_n$.

It should be noted that the effects of external gravitational forces and of interparticle nuclear forces are omitted from our discussion. External gravitational forces can easily be included in the formulation of the momentum transfer equation if necessary. Gravitational forces are often much weaker than electromagnetic forces and thus can often be neglected. If the plasma is so compressed that nuclear forces are significant, one must resort to a quantum-mechanical description of the plasma—the classical methods of analysis used in this book are invalid.

If the sum of the forces on the electron fluid in d^3r is equated to the mass of the electron fluid in d^3r times its acceleration, the resulting equation is

$$\begin{aligned} m_e n_e \, d^3r \frac{d\mathbf{v}_e}{dt} = &- (\nabla P_e) \, d^3r - |e| n_e \, d^3r \, (\mathbf{E} + \mathbf{v}_e \times \mathbf{B}) \\ &- n_e(\alpha_{ei}n_i + \alpha_{en}n_n)\mathbf{v}_e \, d^3r \end{aligned} \tag{4-9}$$

Equation (4-9) can be solved for the acceleration of the electron fluid.

$$\frac{d\mathbf{v}_e}{dt} = -\frac{1}{n_e m_e}\nabla P_e - \frac{|e|}{m_e}(\mathbf{E} + \mathbf{v}_e \times \mathbf{B}) - \nu_{ce}\mathbf{v}_e \tag{4-10}$$

Since collisions cause the damping or frictional force on the electrons, the quantity $[(\alpha_{ei}n_i + \alpha_{en}n_n)/m_e]$ has been defined as the electron *effective* collision frequency ν_{ce}. The dimensions of ν_{ce} are $(\text{time})^{-1}$ or frequency.

The momentum transfer equations for the ions and for the neutral particles can be derived in exactly the same way as the momentum

[1] An alternative assumption as to the functional form of the frictional-force term is given in Appendix B.

transfer equation for the electrons. The results of these derivations are

$$\frac{d\mathbf{v}_i}{dt} = -\frac{1}{n_i m_i}\nabla P_i + \frac{Z_i|e|}{m_i}(\mathbf{E} + \mathbf{v}_i \times \mathbf{B}) - \nu_{ci}\mathbf{v}_i \tag{4-11}$$

$$\frac{d\mathbf{v}_n}{dt} = -\frac{1}{n_n m_n}\nabla P_n - \nu_{cn}\mathbf{v}_n \tag{4-12}$$

where $Z_i|e|$ is the charge which the ion carries. It should be emphasized that the three momentum transfer equations are coupled through the *effective* collision frequencies ν_{ce}, ν_{ci}, and ν_{cn}. We shall have more to say about the effective collision frequencies in Chap. 10.

4-3-3 The Continuity Equations. There are three additional equations describing electron, ion, and neutral-particle behavior which can be derived from the fluid model. These are the continuity equations. If more of a fluid is flowing into a given region in space than is flowing out of that region, there must be a buildup of fluid mass in that region with time. Expressed mathematically, the equations of continuity for electrons, ions, and neutrals, respectively, are

$$\nabla \cdot (n_e\mathbf{v}_e) + \frac{\partial n_e}{\partial t} = 0 \tag{4-13}$$

$$\nabla \cdot (n_i\mathbf{v}_i) + \frac{\partial n_i}{\partial t} = 0 \tag{4-14}$$

$$\nabla \cdot (n_n\mathbf{v}_n) + \frac{\partial n_n}{\partial t} = 0 \tag{4-15}$$

We shall now derive Eq. (4-14). The derivations of Eqs. (4-13) and (4-15) are identical. Consider again a differential volume d^3r of plasma as shown in Fig. 4-2. The x component of ion average velocity

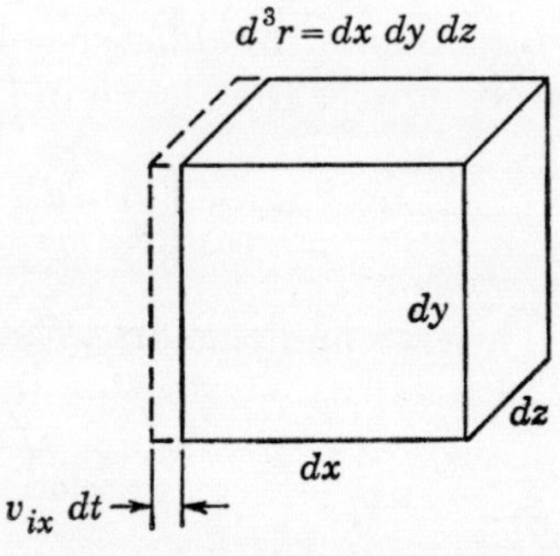

Fig. 4-2 The volume $v_{ix}\,dt\,dy\,dz$ containing the number of ions which in time dt will pass through the left-hand face of d^3r.

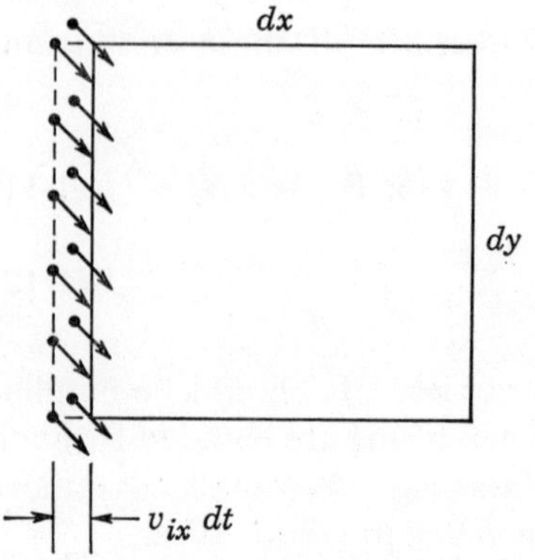

Fig. 4-3 A selected number of ions and the distances these ions travel in time *dt*.

at the left-hand face of d^3r is v_{ix}. The number of ions entering the volume $d^3r = dx\,dy\,dz$ through the left-hand face in time dt can be computed as follows: All ions in the vicinity of the left-hand face can be considered to have the same three components of velocity, the fluid velocity at the left-hand face. Ions whose perpendicular distance to the left-hand face is less than $v_{ix}\,dt$ will pass through the left-hand face in time dt if their y and z coordinates are appropriate. The number of ions that pass through the left-hand face in dt is equal to the total number of ions contained within the dotted volume of Fig. 4-2. This can best be illustrated pictorially in two dimensions. In Fig. 4-3, a selected number of ions and the distances these ions travel in dt are shown. It is apparent that for each ion which begins in the dotted volume and does not pass through the left-hand face, there is another ion which does not begin in the dotted volume and does pass through the left-hand face. The number of ions entering the left-hand face in time dt is therefore $n_i v_{ix}\,dt\,dy\,dz$. The number of ions leaving the right-hand face in dt is given by a Taylor expansion of the number entering the left-hand face, $\{n_i v_{ix} + [\partial(n_i v_{ix})/\partial x]\,dx\}\,dt\,dy\,dz$. The increase in the number of ions in d^3r in time dt due to flow in the x direction is the difference between the number of ions that enter and the number that leave in the x direction,

$$\left[\frac{\partial(n_i\,d^3r)}{\partial t}\right]_x dt = -\left[\frac{\partial(n_i v_{ix})}{\partial x}\right] dt\,d^3r$$

Thus we find that because of flow in the x direction,

$$\left(\frac{\partial n_i}{\partial t}\right)_x = -\frac{\partial(n_i v_{ix})}{\partial x} \tag{4-16}$$

After considering the flow of ions across the other two sets of faces and expressing the result in vector notation, we arrive at the continuity equation for ions

$$\nabla \cdot (n_i \mathbf{v}_i) + \frac{\partial n_i}{\partial t} = 0 \qquad (4\text{-}14)$$

4-3-4 Discussion. We have derived six hydrodynamic equations [Eqs. (4-10) to (4-15)]. An inspection of these equations will show that there are more variables than equations. The variables, excluding the electric and magnetic fields, are n_e, n_i, n_n, $\mathbf{v}_e$, $\mathbf{v}_i$, $\mathbf{v}_n$, P_e, P_i, P_n—nine unknowns, six equations. We need three more equations relating the unknown quantities in order to be able to solve for the nine unknowns. These additional three equations can often be written

$$P_e = n_e \kappa T_e \qquad (4\text{-}17)$$
$$P_i = n_i \kappa T_i \qquad (4\text{-}18)$$
$$P_n = n_n \kappa T_n \qquad (4\text{-}19)$$

where T is the kinetic temperature and κ is the Boltzmann constant. The three equations relating pressure, particle density, and temperature within the plasma are called "equations of state." The particular form of these equations to be used depends on the physics of the particular plasma system under study. The state equations are specified by thermodynamic considerations.

4-4 The Boltzmann Equation

The Boltzmann equation is a mathematical relation which specifies the phase-space distribution function for a particular system of particles (electrons, ions, or neutrals). If a solution to the Boltzmann equation can be found (initial conditions and boundary conditions must be given), all properties of the system under consideration can be calculated from this solution. The Boltzmann equation provides a more detailed description of the behavior of a system of particles than the hydrodynamic equations. The hydrodynamic equations describe the average properties of a system of particles; a solution to the Boltzmann equation reveals how the particles of the system are distributed in velocity and in position.

It is not necessary to use an auxiliary relation with the Boltzmann equation to effect a complete solution, as is the case with the hydrodynamic equations. As the reader has probably guessed, the Boltzmann equation is also much more difficult to solve than the corresponding hydrodynamic equations. We might also expect that the momentum transfer equation and the continuity equation could be derived from the Boltzmann equation. This derivation is given in Appendix A.

Each particle of the plasma is described by a point in six-dimensional phase space. The point in phase space specifies the position and velocity of the particle. Without loss of generality, we can discuss separately a phase space for ions, a phase space for electrons, and a phase space for neutral particles. The ion phase-space distribution function $f_i(\mathbf{r}_i,\mathbf{u}_i,t)$ is equal to the density of representative points in ion phase space; the electron phase-space distribution function $f_e(\mathbf{r}_e,\mathbf{u}_e,t)$ is equal to the density of representative points in electron phase space; and the neutral-particle phase-space distribution function $f_n(\mathbf{r}_n,\mathbf{u}_n,t)$ is equal to the density of representative points in neutral-particle phase space. The Boltzmann equations for the electron phase-space distribution function, for the ion phase-space distribution function, and for the neutral-particle phase-space distribution function can be written

$$\frac{\partial f_e}{\partial t} + \mathbf{u}_e \cdot \nabla_r f_e - \frac{|e|}{m_e} (\mathbf{E} + \mathbf{u}_e \times \mathbf{B}) \cdot \nabla_u f_e = \left(\frac{\partial f_e}{\partial t}\right)_{\text{collisions}} \tag{4-20}$$

$$\frac{\partial f_i}{\partial t} + \mathbf{u}_i \cdot \nabla_r f_i + \frac{Z_i|e|}{m_i} (\mathbf{E} + \mathbf{u}_i \times \mathbf{B}) \cdot \nabla_u f_i = \left(\frac{\partial f_i}{\partial t}\right)_{\text{collisions}} \tag{4-21}$$

$$\frac{\partial f_n}{\partial t} + \mathbf{u}_n \cdot \nabla_r f_n = \left(\frac{\partial f_n}{\partial t}\right)_{\text{collisions}} \tag{4-22}$$

We shall derive the Boltzmann equation for electrons [Eq. (4-20)].

As we have discussed, each electron is represented by a point in six-dimensional electron phase space. The coordinates of the phase space are x, y, z, u_x, u_y, u_z. Because each electron has a velocity, the representative point for each electron must change its position coordinates with time. In addition, any change in the velocity of the electron will also cause the representative point to move in electron phase space. The velocity of an electron will change if a force acts on the electron or if the electron collides with any other particle.

We shall consider a fixed differential volume in electron phase space $d\mathcal{V} = dx\, dy\, dz\, du_x\, du_y\, du_z$ (the "electron" subscript will be omitted). We shall write an expression for the number of representative points in this volume as a function of $\mathbf{r}$, $\mathbf{u}$, and t; that is, we shall write a "continuity" equation for the representative points in electron *phase* space. The increase with time in the number of electron "points" in $d\mathcal{V}$, $\partial(f_e\, d\mathcal{V})/\partial t$ is caused by three factors:

1. The drift of points in position coordinates due to inherent velocity

2. The drift of points in velocity coordinates due to electric and magnetic forces

3. The spontaneous velocity change without position coordinate change due to collisions

When these three terms are evaluated, the Boltzmann equation for electrons can be written down. Note that we again omit the effects of external gravitational forces and interparticle nuclear forces.

Let us first determine the increase with time of the number of representative points in $d\mathcal{V}$ due to drift in position coordinate space. We must calculate the number of representative points which cross the surfaces of $d\mathcal{V}$ in time dt because of inherent velocity. The velocity of a point in the x direction at x, y, z, u_x, u_y, u_z is u_x. The number of representative points in a volume of "base area" $dy\, dz\, du_x\, du_y\, du_z$ and height $u_x\, dt$ will enter $d\mathcal{V}$ in time dt through a face normal to the x axis at x, y, z, u_x, u_y, u_z. This number is

$$\left[\frac{\partial(d^6N)}{\partial t}\right]_x dt = f_e(\mathbf{r},\mathbf{u},t)u_x\, dt\, dy\, dz\, du_x\, du_y\, du_z$$

The change in the number of representative points in $d\mathcal{V}$ in time dt due to the exit of points through the equivalent surface at $x + dx$ is given by

$$\left[\frac{\partial(d^6N)}{\partial t}\right]_{x+dx} dt = -\left[f_eu_x + \frac{\partial(f_eu_x)}{\partial x}\, dx\right] dt\, dy\, dz\, du_x\, du_y\, du_z$$

The net rate of change of points in $d\mathcal{V}$ due to velocity in the x direction is therefore

$$\left[\frac{\partial(d^6N)}{\partial t}\right]_x + \left[\frac{\partial(d^6N)}{\partial t}\right]_{x+dx} = -\frac{\partial(u_xf_e)}{\partial x}\, d\mathcal{V}$$

Thus the total rate of change of points in $d\mathcal{V}$ due to inherent velocity can be written

$$\left[\frac{\partial(d^6N)}{\partial t}\right]_r = -(\mathbf{u}_e \cdot \nabla_r f_e)\, d\mathcal{V} \tag{4-23}$$

We consider now the drift of representative points in velocity coordinates due to electric and magnetic forces. The "velocity" of a representative point in the u_x direction in velocity coordinates is $(du_x/dt) = (F_{ex}/m_e)$. Using this "velocity" we can calculate the net influx of points through the two surfaces perpendicular to u_x. The result of this calculation is

$$\left[\frac{\partial(d^6N)}{\partial t}\right]_{u_x} + \left[\frac{\partial(d^6N)}{\partial t}\right]_{u_x+du_x} = -\left[\frac{\partial}{\partial u_x}\left(\frac{F_{ex}}{m_e} f_e\right)\right] d\mathcal{V}$$

or, if the force in the x direction is independent of u_x,

$$\left[\frac{\partial(d^6N)}{\partial t}\right]_{u_x} + \left[\frac{\partial(d^6N)}{\partial t}\right]_{u_x+du_x} = -\frac{F_{ex}}{m_e}\frac{\partial f_e}{\partial u_x} d\mathcal{V}$$

The extension of this expression to include the flow through the surfaces perpendicular to u_y and u_z follows in an analogous manner. The result is

$$\left[\frac{\partial(d^6N)}{\partial t}\right]_u = -\left(\frac{\mathbf{F}_e}{m_e} \cdot \nabla_u f_e\right) d\mathcal{V} \tag{4-24}$$

At this juncture, we shall equate the time rate of increase of representative points in $d\mathcal{V}$ to the influx from the two drifts and from collisions.

$$\left(\frac{\partial f_e}{\partial t}\right) d\mathcal{V} = -(\mathbf{u}_e \cdot \nabla_r f_e)\, d\mathcal{V} - \left(\frac{\mathbf{F}_e}{m_e} \cdot \nabla_u f_e\right) d\mathcal{V} + \left(\frac{\partial f_e}{\partial t}\right)_{\text{collisions}} d\mathcal{V} \tag{4-25}$$

where $d\mathcal{V}$ is constant in space and in time. If Eq. (4-25) is divided by $d\mathcal{V}$ and rearranged, the resulting equation is the Boltzmann equation for electrons [Eq. (4-20)]:

$$\frac{\partial f_e}{\partial t} + \mathbf{u}_e \cdot \nabla_r f_e + \frac{\mathbf{F}_e}{m_e} \cdot \nabla_u f_e = \left(\frac{\partial f_e}{\partial t}\right)_{\text{collisions}} \tag{4-20}$$

where $\mathbf{F}_e = -|e|(\mathbf{E} + \mathbf{u}_e \times \mathbf{B})$. Of course, the derivation of the Boltzmann equation for ions and the Boltzmann equation for neutral particles is exactly analogous to the foregoing derivation.

We have yet to say anything about the term on the right-hand side of the Boltzmann equation, the collision term. In general, it is the collision term which differentiates the Boltzmann equation for one system of particles from the Boltzmann equation for another system of particles. The calculation of the collision term necessitates a knowledge of the type of collision interaction taking place in the system. The three Boltzmann equations (4-20) to (4-22) are, in general, coupled through their collision terms. For example, the collision terms in the Boltzmann equation for electrons may contain the ion distribution function and the neutral-particle distribution function in addition to the electron distribution function. The collision term is usually quite complex, consisting of integrals over products of distribution functions. Thus the Boltzmann equation is a nonlinear integrodifferential equation, a very difficult type of equation to solve. In many cases plasmas are tenuous enough so that collisions can be ignored and the collision term set equal to zero. We shall have more to say about the collision term in Chap. 10.

4-5 Current Densities and Charge Densities

Since the solution of the hydrodynamic equations or of the Boltzmann equations will be used to calculate the current densities and charge densities which appear in Maxwell's equations, it is appropriate at this point to discuss the calculation of these quantities.

From a hydrodynamic point of view, the charge density ρ and the current density $\mathbf{j}$ are expressed as

$$\rho = (Z_i|e|n_i - |e|n_e) \tag{4-26}$$
$$\mathbf{j} = (Z_i|e|n_i\mathbf{v}_i - |e|n_e\mathbf{v}_e) \tag{4-27}$$

The net charge density is the difference in the charge densities of the ions and the electrons. Each ion carries a charge equal to $Z_i|e|$, and each electron a charge of $-|e|$. The contribution to the current density made by the ions is the charge carried by an ion multiplied by the number of ions which cross an imaginary surface 1 m square perpendicular to the ion flow in 1 sec. The contribution to the current density due to the electron average velocity is opposite to the direction of the electron average velocity.

From the point of view of the distribution-function theory, the charge and current densities are given by

$$\rho = Z_i|e| \iiint_{\text{all } \mathbf{u}} f_i\, d^3u_i - |e| \iiint_{\text{all } \mathbf{u}} f_e\, d^3u_e \tag{4-28}$$

$$\mathbf{j} = Z_i|e| \iiint_{\text{all } \mathbf{u}} \mathbf{u}_i f_i\, d^3u_i - |e| \iiint_{\text{all } \mathbf{u}} \mathbf{u}_e f_e\, d^3u_e \tag{4-29}$$

These expressions can be made to look more familiar by recalling that

$$n = \iiint_{\text{all } \mathbf{u}} f\, d^3u \tag{2-8}$$

$$\mathbf{v} \equiv \bar{\mathbf{u}} = \frac{\iiint_{\text{all } \mathbf{u}} \mathbf{u} f\, d^3u}{\iiint_{\text{all } \mathbf{u}} f\, d^3u} \tag{2-18}$$

Thus Eqs. (4-28) and (4-29) can be rewritten

$$\rho = (Z_i|e|n_i - |e|n_e) \tag{4-30}$$

$$\mathbf{j} = (Z_i|e|n_i\bar{\mathbf{u}}_i - |e|n_e\bar{\mathbf{u}}_e) \tag{4-31}$$

which are identical with Eqs. (4-26) and (4-27), respectively.

PROBLEMS

4-1 In deriving the momentum transfer equation for electrons [Eq. (4-10)], we equated the total force on the electron fluid contained within a small volume d^3r to the mass of the electron fluid within d^3r multiplied by the acceleration of that fluid mass. From a rigorous point of view, Newton's equation of motion states that the force on a system is equal to the rate of change of the *momentum* of the system. In the case of an electron fluid, the momentum of the fluid mass within d^3r is $(n_e m_e \mathbf{v}_e\, d^3r)$. Therefore, if our derivation of the momentum transfer equation is to be valid, the following relation must hold:

$$\frac{d}{dt}(n_e m_e \mathbf{v}_e\, d^3r) = n_e m_e\, d^3r\, \frac{d\mathbf{v}_e}{dt}$$

Remembering that we have written Newton's equation of motion for a system whose mass remains constant and that n_e is a function of position and time, prove that the above equation is valid. What can be concluded about the variation in the size of d^3r with time? Does the size of the differential volume d^3r used in the derivation of the continuity equation for the electron fluid vary with time?

4-2 List all assumptions made in deriving the momentum transfer equations and the continuity equations.

4-3 List all assumptions made in deriving the Boltzmann equations.

4-4 Discuss the following statement: The Boltzmann equation is not a valid mathematical representation of a system of particles in which collisions are so frequent that the trajectory of one particle may no longer be considered to be composed of relatively long segments during which only external forces act joined by very short segments during which there is a collision.

4-5 Discuss the following statement: The Boltzmann equation is not a valid mathematical representation of a system of particles in which the forces between particles decrease very slowly with distance.

4-6 The following vector relationship was used in deriving Eq. (4-24):

$$\frac{\mathbf{F}_e}{m_e} \cdot \nabla_u f_e = \nabla_u \cdot \frac{\mathbf{F}_e}{m_e} f_e$$

Show that this relationship is valid if

$$\mathbf{F}_e = -|e|(\mathbf{E} + \mathbf{u}_e \times \mathbf{B})$$

4-7 Formulate the momentum transfer equation for the neutral particles of a plasma if the plasma is immersed in a gravitational field $\mathbf{G}(\mathbf{r})$ newtons/kilogram.

4-8 To obtain an idea of the relative importance of interparticle gravitational forces versus interparticle electric forces, calculate the ratio of the gravitational force of attraction between an electron and an oxygen ion to the Coulomb force of attraction between these two particles. How does this ratio vary with interparticle separation?

4-9 The continuity equation for charge is

$$\frac{\partial \rho}{\partial t} + \nabla \cdot \mathbf{j} = 0$$

Obtain this equation in two ways: (*a*) by taking the curl of both sides of Eq. (4-2*a*) and combining the resultant expression with Eq. (4-1*a*) and (*b*) by combining Eqs. (4-13) and (4-14) and making use of Eqs. (4-26) and (4-27).

4-10 The continuity equation for a system of particles in real space is

$$\frac{\partial n}{\partial t} + \nabla \cdot (n\mathbf{v}) = 0$$

Show that in the absence of collisions the Boltzmann equation can be cast into the form of a continuity equation for representative points in six-dimensional phase space

$$\frac{\partial f}{\partial t} + \nabla_r \cdot (f\mathbf{u}) + \nabla_u \cdot (f\mathbf{a}) = 0$$

where $\mathbf{a} = \mathbf{F}/m$ is the acceleration.

BIBLIOGRAPHY

Binder, R. C.: "Fluid Mechanics," 4th ed., Prentice-Hall, Inc., Englewood Cliffs, N.J., 1962.
chap. 20—On the derivation of the equations of motion for a fluid using a technique similar to that given within the chapter.

Brueckner, K. A., and K. M. Watson: Use of the Boltzmann Equation for the Study of Ionized Gases of Low Density, II, *Phys. Rev.*, **102**:19 (1956).

Chapman, S., and T. G. Cowling: "The Mathematical Theory of Non-uniform Gases," Cambridge University Press, New York, 1953.
p. 46—On the derivation of the Boltzmann equation.

Chew, G. F., M. L. Goldberger, and F. E. Low: The Boltzmann Equation and the One-fluid Hydromagnetic Equations in the Absence of Particle Collisions, *Proc. Roy. Soc. (London)*, **236**:112 (1956).

Hayt, W. H., Jr.: "Engineering Electromagnetics," McGraw-Hill Book Company, Inc., New York, 1958.
p. 249—On Maxwell's equations in differential and in integral form.

Oster, L.: Linearized Theory of Plasma Oscillations, *Rev. Mod. Phys.*, **32**:141 (1960).
On the hydrodynamic versus the Boltzmann equation approach to plasma physics.

Rosenbluth, M. N., and N. Rostoker: Theoretical Structure of Plasma Equations, *Phys. Fluids*, **2**:23 (1959).
On the application of the equations of plasma physics to the case of a collisionless plasma.

Spitzer, L., Jr.: "Physics of Fully Ionized Gases," 2d ed., Interscience Publishers, Inc., New York, 1962.
appendix—On the derivation of the hydrodynamic equations from the Boltzmann equation.

Watson, K. M.: Use of the Boltzmann Equation for the Study of Ionized Gases of Low Density, I, *Phys. Rev.*, **102**:12 (1956).

chapter

5

The Behavior of Charged Particles in Electric and Magnetic Fields

5-1 Introduction

The behavior of a plasma is determined by considering the interaction of the plasma particles with themselves and with externally applied electric and magnetic fields. In this chapter we shall investigate the behavior of the charged plasma particles as they interact with externally applied electric and magnetic fields.

In the previous chapters we introduced two velocity variables: **u** and **v**. The velocity variable **u** represents the total velocity of an individual plasma particle; the velocity variable **v** represents the average velocity of a system of particles whose phase-space distribution function is $f(\mathbf{u},\mathbf{r},t)$.

$$\mathbf{v}(\mathbf{r},t) = \frac{\iiint\limits_{\text{all } \mathbf{u}} \mathbf{u} f(\mathbf{u},\mathbf{r},t)\, d^3u}{\iiint\limits_{\text{all } \mathbf{u}} f(\mathbf{u},\mathbf{r},t)\, d^3u} \qquad (2\text{-}18)$$

We shall use the velocity variable **u** when we consider the motion of a single plasma particle, and the velocity variable **v** when we consider the fluidlike behavior of one group of identical plasma particles.

In the absence of collisions, the equation which specifies the motion of a single charged particle is

$$m\frac{d\mathbf{u}}{dt} = q(\mathbf{E} + \mathbf{u} \times \mathbf{B}) \qquad (5\text{-}1)$$

If collisions occur within a plasma, that plasma may be analyzed from a fluid viewpoint. The equation which specifies the behavior of one of the three intermingled plasma fluids is the momentum transfer equation

$$m\frac{d\mathbf{v}}{dt} = -\frac{\nabla P}{n} + q(\mathbf{E} + \mathbf{v} \times \mathbf{B}) - m\nu_c\mathbf{v} \qquad (5\text{-}2)$$

Equations (5-1) and (5-2) can be applied either to the plasma electrons ($q = -|e|$, e subscripts) or to the plasma ions ($q = Z_i|e|$, i subscripts).

In this chapter we shall examine several cases of plasma-field interactions for which collisions between plasma particles are considered negligible and several cases of plasma-field interactions for which collisions between plasma particles are included by postulating a *constant* effective collision frequency. In Sec. 5-2 the behavior of individual charged particles in various configurations of electric and magnetic fields is investigated. In this study we shall assume that the plasma particles do not suffer collisions. Thus we shall concern ourselves with the solution of Eq. (5-1). In Sec. 5-3 we consider the fluidlike behavior of a plasma. Assuming that for the system of particles under study the effective collision frequency is constant and the pressure gradient is zero, we shall solve the momentum transfer equation, Eq. (5-2).

The results obtained in this chapter will be useful from two points of view: (1) They will specify the plasma behavior for the field configurations which we shall analyze, and (2) they will give us a physical feeling for the behavior of a plasma in field configurations too complex to be conveniently analyzed mathematically.

5-2 The Motion of Charged Particles in the Absence of Collisions

5-2-1 Static Uniform Electric Field. The equation which specifies the motion of a charged particle in a uniform electrostatic field is

$$m\frac{d\mathbf{u}}{dt} = q\mathbf{E} \qquad (5\text{-}3)$$

We shall calculate the trajectory of a charged particle which at time $t = 0$ finds itself in an electric field $\mathbf{E}$ in the y direction as shown in Fig. 5-1. At $t = 0$ the three velocity components of the particle are assumed to be u_{0x}, u_{0y}, u_{0z}. Equation (5-3) can be written in com-

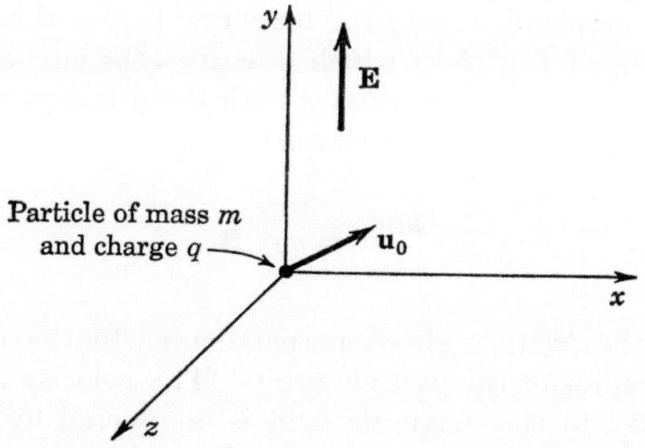

Fig. 5-1 A charged particle in a static uniform electric field at $t = 0$.

ponent form as follows:

$$m\frac{du_x}{dt} = 0 \qquad m\frac{du_y}{dt} = qE \qquad m\frac{du_z}{dt} = 0 \tag{5-4}$$

Equation (5-4) can be solved by direct integration. These solutions, with initial conditions included, are

$$u_x = u_{0x} \qquad u_y = u_{0y} + \frac{qEt}{m} \qquad u_z = u_{0z} \tag{5-5}$$

The motion of the charged particle in the direction perpendicular to the electric field is unaffected. The additional particle velocity caused by the field increases linearly with time in the direction of the electric field if the particle is positively charged and in the direction opposite the electric field if the particle is negatively charged. An electrostatic field applied to a plasma will cause charge separation since the ions and electrons will be forced in opposite directions by the field.

5-2-2 Static Uniform Magnetic Field. The equation which specifies the motion of a charged particle in a uniform magnetostatic field is

$$m\frac{d\mathbf{u}}{dt} = q(\mathbf{u} \times \mathbf{B}) \tag{5-6}$$

The acceleration appearing on the left-hand side of Eq. (5-6) is perpendicular to both the velocity and the magnetic field. The particle trajectory perpendicular to the magnetic field is therefore a circle. The component of Eq. (5-6) which describes the circular motion can be written in cylindrical coordinates with the origin of the coordinate system at the center of the circular orbit,

$$\frac{mu_\perp^2}{r_b} = qu_\perp|\mathbf{B}| \tag{5-7}$$

where $u_\perp$ is the particle speed perpendicular to the magnetic field and r_b is the radius of the particle orbit. The velocity of the charged particle parallel to the magnetic field is unaffected by the magnetic field. The total trajectory of the particle is a helix. The angular frequency with which the particle orbits the magnetic field can be found from Eq. (5-7).

$$\omega_b = \frac{u_\perp}{r_b} = \frac{q|\mathbf{B}|}{m} \tag{5-8}$$

A diagram of the particle orbit is shown in Fig. 5-2. The orbital angular frequency will be called the "cyclotron frequency." We can find the time taken for one circular turn τ_b by multiplying the reciprocal of the cyclotron frequency by 2π.

$$\tau_b = \frac{2\pi m}{q|\mathbf{B}|} \tag{5-9}$$

It is worth noting that the period of the particle motion τ_b is independent of the particle speed. The constancy of the period is the basis of operation of the cyclotron.

A parameter which will be of much interest to us is the magnetic dipole moment associated with an orbiting charged particle. The magnetic dipole moment $\mathcal{U}$ is defined as the product of the current flowing around the orbit and the area enclosed by the orbit, $\mathcal{U} = I\pi r_b^2$. The current flowing around the orbit is q/τ_b, the charge multiplied by the number of times per second the charge passes a given point on the orbit. The magnetic moment can therefore be written

$$\mathcal{U} = \frac{q\pi r_b^2}{\tau_b} = \frac{(mu_\perp^2)/2}{|\mathbf{B}|} = \frac{\mathcal{E}_\perp}{|\mathbf{B}|} \tag{5-10}$$

where $\mathcal{E}_\perp$ is the energy of the particle associated with the particle motion perpendicular to the magnetic field.

The magnetic flux φ passing through the particle orbit is found by

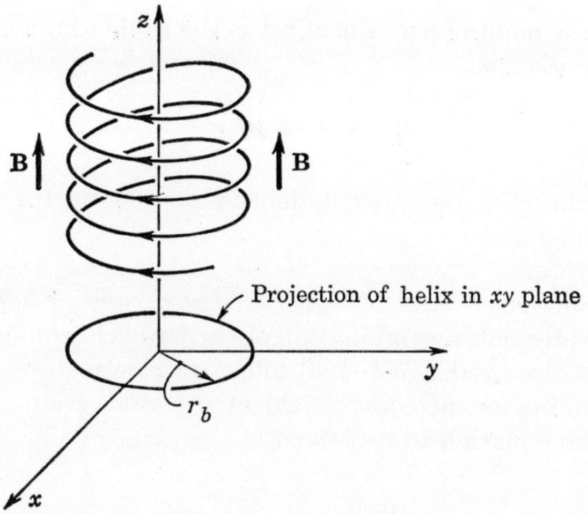

Fig. 5-2 Helical trajectory of a positively charged particle in a static uniform magnetic field.

multiplying the flux density by the area of the orbit. The magnetic flux through the particle orbit is found to be related to the magnetic moment by a constant

$$\varphi = |\mathbf{B}|\pi r_b^2 = \frac{2\pi m}{q^2}\,\mathfrak{U} \tag{5-11}$$

If there are a large number of identical plasma particles orbiting in a magnetic field, there will, in general, be a range of magnetic moments present. That there is a distribution of magnetic moments follows from the fact that there is, in general, a distribution of particle energies. The magnetic moment per unit volume of the identical particles is defined as the *average* magnetic moment density of the orbiting particles. The *vector* magnetic moment per unit volume **M** due to both the ions and the electrons of a plasma is

$$\mathbf{M}(\mathbf{r}) = -n_i\overline{\mathfrak{U}_i}\,\frac{\mathbf{B}}{|\mathbf{B}|} - n_e\overline{\mathfrak{U}_e}\,\frac{\mathbf{B}}{|\mathbf{B}|} \tag{5-12}$$

where $\overline{\mathfrak{U}} = \overline{\mathcal{E}_\perp}/|\mathbf{B}|$ and where the direction associated with the dipole moments is opposite the direction of the magnetic field; that is, the plasma is diamagnetic. If the magnetic moment per unit volume of the plasma is nonuniform, the effect is the same as if a current were flowing in the plasma.

$$\mathbf{j}_M = \nabla \times \mathbf{M}(\mathbf{r}) \tag{5-13}$$

where $\mathbf{j}_M$ is the effective current density caused by the nonuniform magnetic moments.

5-2-3 Perpendicular Static Uniform Electric and Magnetic Fields. We shall consider now a combination of the two field configurations we have already discussed. We shall allow an electrostatic field in the y direction and a magnetostatic field in the z direction, as shown in Fig. 5-3. The equation to be solved is

$$m\frac{d\mathbf{u}}{dt} = q(\mathbf{E} + \mathbf{u} \times \mathbf{B}) \tag{5-14}$$

Before obtaining a solution to Eq. (5-14), we shall look physically at what we would expect for the trajectory of the particle. If we assume that a positively charged particle is released from rest at the origin, the initial motion of the particle will be in the direction of the electric field. As the particle gains velocity in the y direction, it will feel a magnetic force in the direction $\mathbf{u} \times \mathbf{B}$ and therefore will move in the x direction. The particle will be forced by the magnetic field to return to the positive x axis, at which time the velocity of the particles will again be zero. The trajectory of the motion will be cycloidal. The cycloidal loop will be repeated over and over, as shown in Fig. 5-4. If the initial y velocity of the positive particle is not zero, the trajectory will be a trochoid. A trochoid is the curve described by a point on a spoke (or on an extension of a spoke) of a wheel as the

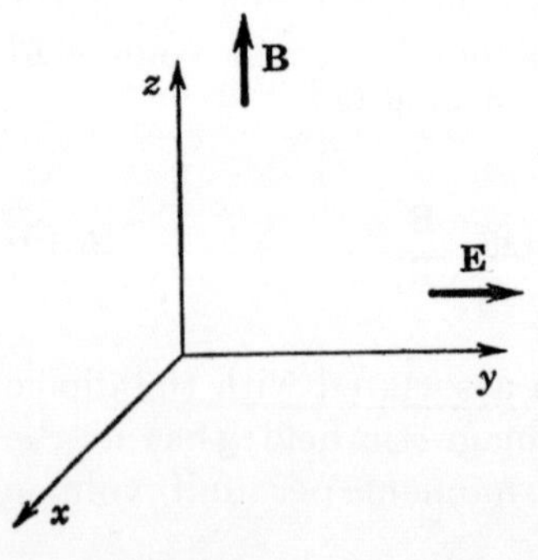

Fig. 5-3 The static uniform fields considered in Sec. 5-2-3.

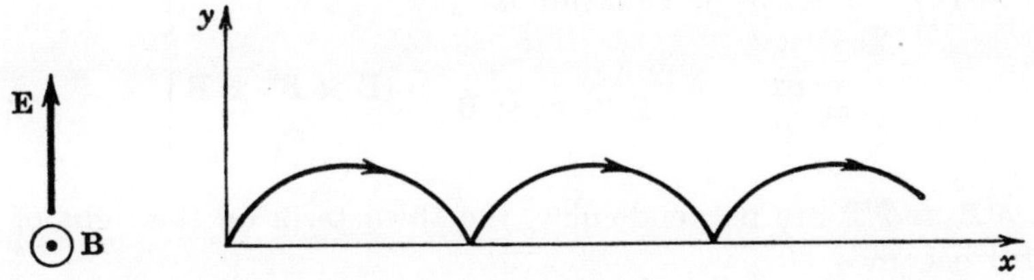

Fig. 5-4 The cycloidal trajectory of a positively charged particle released from rest in the presence of perpendicular electric and magnetic fields.

wheel rolls along a horizontal straight line without slipping. When the tracing point is on the circumference of the wheel, the trochoid is called a cycloid. Figure 5-5 depicts the trajectory of the particle if it is given an initial velocity in the y direction.

We shall now show that the motion of the plasma particle in the crossed electric and magnetic fields is composed of a helical motion superimposed upon a drift of constant velocity in the direction perpendicular to both fields. We shall assume the following solution for Eq. (5-14):

$$\mathbf{u} = \mathbf{u}_b + \frac{\mathbf{E} \times \mathbf{B}}{|\mathbf{B}|^2} \tag{5-15}$$

where the second term on the right of Eq. (5-15) is the constant drift velocity of magnitude $E/|\mathbf{B}|$. If we can find a $\mathbf{u}_b$ such that the assumed velocity variation satisfies Eq. (5-14), Eq. (5-15) is a valid representation for the velocity of the particle. If we substitute Eq. (5-15) into

Fig. 5-5 The trochoidal trajectory of a positively charged particle released with an initial velocity in the y direction in the presence of perpendicular electric and magnetic fields.

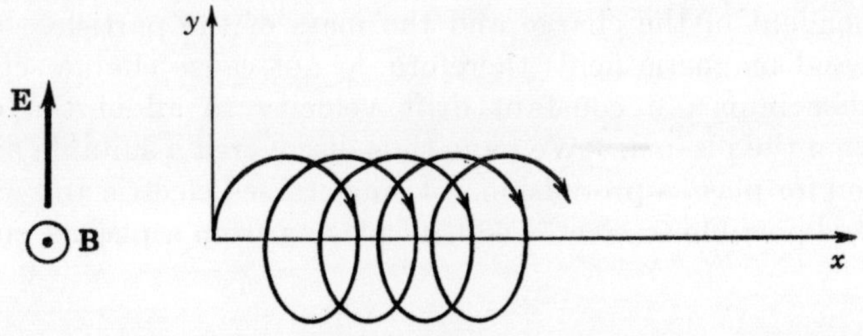

Eq. (5-14), the resulting equation is

$$m\frac{d\mathbf{u}_b}{dt} = q\left[\mathbf{E} + \mathbf{u}_b \times \mathbf{B} + \frac{(\mathbf{E}\times\mathbf{B})\times\mathbf{B}}{|\mathbf{B}|^2}\right] \tag{5-16}$$

Since **E** and **B** are perpendicular, the third term on the right of Eq. (5-16) becomes

$$\frac{(\mathbf{E}\times\mathbf{B})\times\mathbf{B}}{|\mathbf{B}|^2} = -\mathbf{E}$$

Equation (5-16) thus becomes

$$m\frac{d\mathbf{u}_b}{dt} = q(\mathbf{u}_b \times \mathbf{B}) \tag{5-17}$$

which is identical with Eq. (5-6). The magnetic moment of the charged particle therefore remains constant as the particle moves parallel to the magnetic field with a velocity determined by initial conditions and perpendicular to both the electric field and the magnetic field with a drift velocity

$$\mathbf{v}_E = \frac{\mathbf{E}\times\mathbf{B}}{|\mathbf{B}|^2} \tag{5-18}$$

Notice that we have used the velocity variable $\mathbf{v}$ rather than $\mathbf{u}$ in Eq. (5-18). The reason for this choice of notation is as follows: If a collisionless plasma is immersed in perpendicular electrostatic and magnetostatic fields, all members of each system of identical plasma particles move with the same drift velocity $\mathbf{v}_E$ regardless of their orbital velocities. The phases of the orbital motion of the various identical particles will in general be random, and thus the orbital velocities can yield no average velocity. Equation (5-18) therefore represents a fluid velocity or average velocity of the system of identical plasma particles. The use of the velocity variable $\mathbf{v}$ in Eq. (5-18) is therefore consistent with our previous notation.

The direction of the charged-particle drift as given by Eq. (5-18) is independent of the charge and the mass of the particle. Crossed electric and magnetic fields therefore do not cause charge separation but rather impart a constant drift velocity to all of the charged particles in the plasma. We have thus discovered a suitable field configuration for plasma propulsion. Using crossed electric and magnetic fields, it is possible to eject a neutral plasma from a plasma engine.

5-2-4 The Motion of Guiding Centers. We have seen in our discussion of particle motion in crossed electric and magnetic fields that the exact trajectory of a charged particle may be decomposed into an orbital motion and a drift velocity. The drift velocity can be considered to be the velocity of the center of the orbital motion. We shall call this velocity the "velocity of the guiding center" of the particle. In this section we shall derive a general expression for the drift velocity of guiding centers in the presence of a uniform static magnetic field and a uniform static external force.

The equation which specifies the behavior of a charged particle in the presence of a uniform static magnetic field and a uniform static force **F** is

$$m\frac{d\mathbf{u}}{dt} = \mathbf{F} + q(\mathbf{u} \times \mathbf{B}) \tag{5-19}$$

Equation (5-19) can be resolved into an equation of motion perpendicular to **B** and an equation of motion parallel to **B**.

$$m\frac{d\mathbf{u}_\perp}{dt} = \mathbf{F}_\perp + q(\mathbf{u}_\perp \times \mathbf{B}) \tag{5-20}$$

$$m\frac{d\mathbf{u}_\parallel}{dt} = \mathbf{F}_\parallel \tag{5-21}$$

where the $\perp$ subscript means "perpendicular to **B**" and the $\parallel$ subscript means "parallel to **B**." The solution to Eq. (5-21) is

$$\mathbf{u}_\parallel = \frac{\mathbf{F}_\parallel}{m}t + \mathbf{u}_{0\parallel} \tag{5-22}$$

where $\mathbf{u}_{0\parallel}$ is the velocity parallel to the magnetic field at $t = 0$. The motion in the direction of the magnetic field is unaffected by the magnetic field.

We shall assume a solution for Eq. (5-20) and then prove that the assumed solution is valid. We shall assume that

$$\mathbf{u}_\perp = \mathbf{u}_{b\perp} + \frac{\mathbf{F}_\perp \times \mathbf{B}}{q|\mathbf{B}|^2} \tag{5-23}$$

Substitution of Eq. (5-23) into Eq. (5-20) yields

$$m\frac{d\mathbf{u}_{b\perp}}{dt} = \mathbf{F}_\perp + q(\mathbf{u}_{b\perp} \times \mathbf{B}) + \frac{(\mathbf{F}_\perp \times \mathbf{B}) \times \mathbf{B}}{|\mathbf{B}|^2} \tag{5-24}$$

Since $\mathbf{F}_\perp$ and $\mathbf{B}$ are perpendicular,

$$\frac{(\mathbf{F}_\perp \times \mathbf{B}) \times \mathbf{B}}{|\mathbf{B}|^2} = -\mathbf{F}_\perp$$

and Eq. (5-24) becomes

$$m \frac{d\mathbf{u}_{b\perp}}{dt} = q(\mathbf{u}_{b\perp} \times \mathbf{B}) \qquad (5\text{-}25)$$

Equation (5-25) is the expression for the orbital motion of a charged particle at the cyclotron frequency. The orbital motion at velocity $\mathbf{u}_{b\perp}$ is unaffected by the presence of $\mathbf{F}_\perp$. The magnetic moment of the charged particle must therefore remain constant. The interaction of $\mathbf{F}_\perp$ and $\mathbf{B}$ causes the guiding center of the orbital motion to drift in a direction perpendicular to both $\mathbf{F}_\perp$ and $\mathbf{B}$. The guiding-center drift is independent of $\mathbf{u}_{b\perp}$. The particle motion may be considered that of a fixed dipole moment whose guiding center drifts with a velocity

$$\mathbf{v}_D = \frac{\mathbf{F}_\perp \times \mathbf{B}}{q|\mathbf{B}|^2} \qquad (5\text{-}26)$$

Notice that we have used the velocity variable $\mathbf{v}$ to designate the drift velocity of the guiding center of the particle. In the absence of collisions, all of the identical plasma particles which are immersed in $\mathbf{F}_\perp$ and $\mathbf{B}$ will move with the drift velocity given by Eq. (5-26) regardless of their orbital velocities. The system of identical particles is represented by the average or fluid velocity given by Eq. (5-26) since the orbital motions are, in general, random in phase and thus cannot contribute to the average velocity.

To check Eq. (5-26), we shall assume that the applied force is caused by an electric field, $\mathbf{F}_\perp = q\mathbf{E}$. Equation (5-26) then becomes

$$\mathbf{v}_E = \frac{\mathbf{E} \times \mathbf{B}}{|\mathbf{B}|^2} \qquad (5\text{-}27)$$

the result we obtained in the previous section. As an additional example we shall consider the applied force to be that of the earth's gravitational field, $\mathbf{F}_\perp = m\mathbf{g}$. Equation (5-26) becomes

$$\mathbf{v}_g = \frac{m\mathbf{g} \times \mathbf{B}}{q|\mathbf{B}|^2} \qquad (5\text{-}28)$$

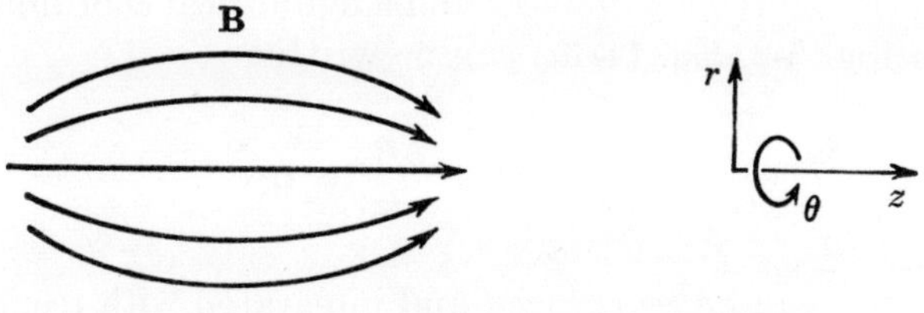

Fig. 5-6 The magnetic mirror field.

The particle drift caused by crossed magnetic and gravitational fields is in opposite directions for oppositely charged particles. The drift velocity of the heavier ions is more rapid than the drift velocity of the electrons. The direction of drift is perpendicular to the direction of both the magnetic field and the gravitational field.

5-2-5 Nonuniform Magnetic Fields

General. We shall discuss a very special type of nonuniform magnetic field, a static uniform field in the z direction B_{0z} with small perturbational fields B_x, B_y, B_z superimposed.[1] This particular field configuration is of considerable importance in the controlled-fusion program. Large static magnetic fields are often used to contain the high-energy plasma. In practice it is impossible to create a perfectly uniform magnetic field. There will therefore always be perturbation fields. The unwanted perturbation fields may cause drifts of particles leading to the destruction of the containment properties of the magnetic "bottle." On the other side of the coin, the perturbation fields may be manipulated so that they enhance containment.

The Magnetic Mirror. $\partial B_z/\partial z$, $\partial B_y/\partial y$, $\partial B_x/\partial x$. A picture of magnetic field lines for which the only important variation of the perturbational fields is $\partial B_z/\partial z$, $\partial B_y/\partial y$, $\partial B_x/\partial x$ is shown in Fig. 5-6. This magnetic field configuration is called a "magnetic mirror" for reasons which will become apparent later.

The magnetic field variables are related through Eq. (4-4*a*),

$$\nabla \cdot \mathbf{B} = 0$$

[1] As before, the total magnetic field is designated **B**, the magnitude of the total magnetic field $|\mathbf{B}|$. Thus for the nonuniform fields that we shall discuss, the magnitude of **B** can be written

$$|\mathbf{B}| = [(B_{0z} + B_z)^2 + B_y^2 + B_x^2]^{1/2}$$

If we assume cylindrical symmetry and a cylindrical coordinate system as shown in Fig. 5-6, Eq. (4-4*a*) can be written

$$\frac{1}{r}\frac{\partial}{\partial r}(rB_r) + \frac{\partial B_z}{\partial z} = 0 \tag{5-29}$$

Equation (5-29) can be rearranged and integrated with respect to r.

$$\int_{r'=0}^{r'=r} \frac{\partial}{\partial r'}(r'B_{r'})\,dr' = -\int_{r'=0}^{r'=r} r'\frac{\partial B_z}{\partial z}\,dr' \tag{5-30}$$

Since the only important field variation is assumed to be $\partial B_z/\partial z$, $\partial B_y/\partial y$, and $\partial B_x/\partial x$, then $\partial B_z/\partial z$ can be considered independent of r, and B_r can be considered only a function of r. Equation (5-30) can be integrated to yield

$$B_r = -\tfrac{1}{2}r\frac{dB_z}{dz} \tag{5-31}$$

where the partial derivative has been replaced by a total derivative since B_z is assumed to be only a function of z.

The force in the z direction on a charged particle orbiting about B_{0z} is

$$F_z = qu_\perp B_r \tag{5-32}$$

where $u_\perp$ is the magnitude of the velocity in the negative θ direction. The combination of Eqs. (5-31) and (5-32) yields

$$F_z = -\tfrac{1}{2}qu_\perp r\frac{dB_z}{dz} \tag{5-33}$$

Using the relations $u_\perp = r\omega_b$, $\omega_b = 2\pi/\tau_b$, $\mathcal{U} = (q\pi r^2)/\tau_b$ and letting $r = r_b$, we can convert Eq. (5-33) into

$$F_z = -\mathcal{U}\frac{dB_z}{dz} \tag{5-34}$$

The force on the charged particle is toward the direction of the weaker magnetic field. Thus the magnetic field shown in Fig. 5-6 acts as a "bottle." As the particle approaches the strong magnetic field at either end of the "mirror" geometry, it is forced back toward the center of the bottle. The ends of the magnetic mirror reflect the particle. The result is particle containment.

When the particle has moved a distance Δz under the influence of the force in the z direction, the particle must gain an energy

$$\Delta \mathcal{E}_{\|} = F_z \, \Delta z = -\mathfrak{M} \frac{dB_z}{dz} \Delta z \tag{5-35}$$

The total energy of a particle in a static magnetic field cannot change. The velocity of the particle is always perpendicular to the force on the particle, and therefore the force can do no work on the particle. The change of particle energy due to motion perpendicular to the z direction therefore must be equal and of opposite sign to the change of particle energy due to motion parallel to the z direction.

$$\Delta \mathcal{E}_{\perp} = \mathfrak{M} \frac{dB_z}{dz} \Delta z \tag{5-36}$$

It follows that

$$\frac{d\mathcal{E}_{\perp}}{dz} = \mathfrak{M} \frac{dB_z}{dz} \tag{5-37}$$

We know, however, from Eq. (5-10) that $\mathcal{E}_{\perp} = \mathfrak{M}(B_{0z} + B_z)$. Thus, in general, the derivative of $\mathcal{E}_{\perp}$ with respect to z must yield

$$\frac{d\mathcal{E}_{\perp}}{dz} = \mathfrak{M} \frac{dB_z}{dz} + (B_{0z} + B_z) \frac{d\mathfrak{M}}{dz} \tag{5-38}$$

After comparing Eq. (5-37) with Eq. (5-38), we must conclude that for the magnetic mirror geometry

$$\frac{d\mathfrak{M}}{dz} = 0$$

or that the magnetic moment is independent of z. Using Eq. (5-11), we can alternatively say that the magnetic flux through the particle orbit must remain constant. As the particle drifts into the region of stronger magnetic field, the orbit of the particle must become smaller so that the flux through the orbit will remain constant.

We shall assume that initially the particle in the magnetic mirror geometry has a velocity component in the z direction u_z in addition to its orbital velocity component $u_{\perp}$. The initial velocity $\mathbf{u}_0$ makes an angle χ_0 with the z axis so that

$$u_{0z} = u_0 \cos \chi_0 \qquad u_{0\perp} = u_0 \sin \chi_0$$

We know that the magnetic moment $\mathfrak{M}$ will remain constant during the particle motion. Initially the magnetic moment is given by

$$\mathcal{U} = \frac{(mu_{0\perp}^2)/2}{B_{az}} \tag{5-39}$$

where B_{az} is the z component of the magnetic field at the point where the particle starts. The magnetic moment at all times must be equal to the value given in Eq. (5-39).

$$\frac{(mu_{0\perp}^2)/2}{B_{az}} = \frac{(mu_{\perp}^2)/2}{B_{bz}} \tag{5-40}$$

where B_{bz} is the z component of the field at the points for which the particle has a perpendicular velocity component $u_\perp$. The total speed of the particle u_0 cannot change, so that we can write $u_\perp = u_0 \sin \chi$, where χ is the angle between the total velocity vector and the z axis. Equation (5-40) can therefore be written

$$\sin^2 \chi = \left(\frac{B_{bz}}{B_{az}}\right) \sin^2 \chi_0 \tag{5-41}$$

We shall assume that the particle is initially moving in the direction of increasing B_{bz}. As B_{bz} increases, the angle χ becomes larger. When the right-hand side of Eq. (5-41) equals unity, the angle χ is equal to 90°. The particle has stopped moving in the z direction. Since there is a force on the particle pushing it toward the weaker field, the particle will move in the direction opposite to its original motion. The particle has been reflected. Both ends of the magnetic field geometry shown in Fig. 5-6 have strong fields and will act as magnetic mirrors. The particle will oscillate back and forth within the magnetic bottle.

We have discussed the motion of an individual particle of magnetic moment $\mathcal{U}$. That motion is determined by Eq. (5-34), in which $\mathcal{U}$ is a constant. If we consider a collisionless plasma immersed in a magnetic mirror field, there will, in general, be a range of values for the magnetic moments of the individual particles because of the spread in energy of the particles. The average force on the plasma electrons or on the plasma ions as a function of position coordinates and of time due to the magnetic mirror field can be found by averaging F_z from Eq. (5-34) over the appropriate distribution function. The result of this manipulation is

$$\overline{F_z(\mathbf{r},t)} = -\overline{\mathcal{U}} \frac{dB_z}{dz} \tag{5-42}$$

where $\overline{\mathcal{U}} = \overline{\mathcal{E}_\perp}/|\mathbf{B}|$ and

$$\overline{\mathcal{E}_\perp(\mathbf{r},t)} = \frac{\iiint\limits_{\text{all } \mathbf{u}} \tfrac{1}{2}m(u_x^2 + u_y^2)f(\mathbf{u},\mathbf{r},t)\, d^3u}{\iiint\limits_{\text{all } \mathbf{u}} f(\mathbf{u},\mathbf{r},t)\, d^3u}$$

Magnetic Field Gradient. $\partial B_z/\partial x$, $\partial B_z/\partial y$. We shall consider now those magnetic field configurations in which a magnetic field in the z direction varies only as a function of x and y. We can qualitatively determine the trajectory of a charged particle under such conditions as follows: For a particle in a uniform magnetic field, the radius of curvature of the orbital motion can be found from Eq. (5-7).

$$\mathbf{r}_b = \frac{\mathbf{p} \times \mathbf{B}}{q|\mathbf{B}|^2} \tag{5-43}$$

where $\mathbf{p} = m\mathbf{u}$, and vector notation has been added to give direction to the radius of curvature. If the magnetic field is not uniform in space, Eq. (5-43) defines the instantaneous radius of curvature of the particle trajectory. The radius of curvature which the particle exhibits is inversely proportional to the magnetic field strength. The behavior of ions and electrons with and without a magnetic field gradient perpendicular to B_{0z} present is shown in Fig. 5-7. As the particle moves into the stronger magnetic field, its radius of curvature becomes smaller. In the weaker field, its radius of curvature becomes larger. The resultant motion represents a drift across the magnetic field lines. The electrons and the ions drift in opposite directions. All particles drift in directions perpendicular to both the magnetic field and the gradient of the magnetic field.

We shall assume that $\partial B_z/\partial x > 0$ and that $\partial B_z/\partial y = 0$, as shown in Fig. 5-7. The integral over one orbital period of the force in the x direction on the charged particle must be equal to zero since the only net force acting on the particle is in the y direction.

$$\int_{\tau_b} F_x\, dt = 0 \tag{5-44}$$

where $F_x = qu_y(B_{0z} + B_z)$. If we assume that the magnetic field varies only slightly over the particle trajectory, the magnetic field in which the particle moves can be written

$$\left[|\mathbf{B}(x_0)| + \frac{\partial B_z}{\partial x}\bigg|_{x_0} (x - x_0) \right]$$

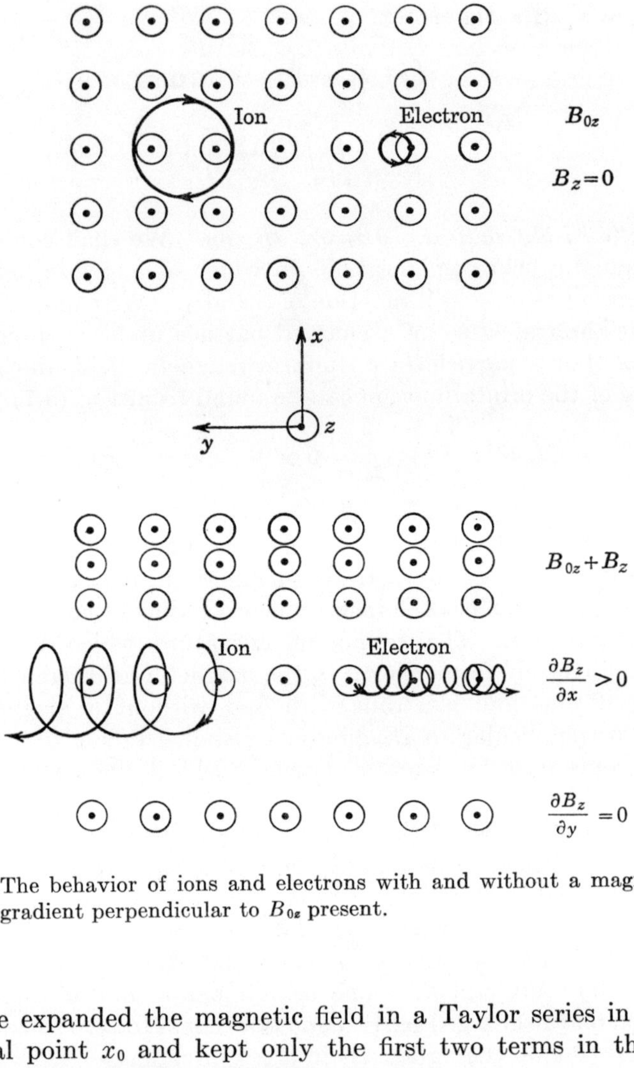

Fig. 5-7 The behavior of ions and electrons with and without a magnetic field gradient perpendicular to B_{0z} present.

We have expanded the magnetic field in a Taylor series in x about a general point x_0 and kept only the first two terms in the series. Equation (5-44) therefore becomes

$$\int_{\tau_b} u_y\,dt = -\int_{\tau_b} \frac{u_y(x - x_0)}{|\mathbf{B}(x_0)|}\left(\frac{\partial B_z}{\partial x}\bigg|_{x_0}\right) dt \tag{5-45}$$

The left-hand side of Eq. (5-45) is the displacement of the particle in the y direction during one orbit.

$$\Delta y = -\frac{1}{|\mathbf{B}(x_0)|}\left(\frac{\partial B_z}{\partial x}\bigg|_{x_0}\right)\int_{\tau_b}(x - x_0)\,dy \tag{5-46}$$

where $u_y\,dt = dy$. If the particle drift is small, the orbits are approximately closed, and for a positive particle the integral in Eq. (5-46) is equal to the negative of the area enclosed within an orbit.

$$\Delta y = \frac{1}{|\mathbf{B}(x_0)|}\left(\frac{\partial B_z}{\partial x}\bigg|_{x_0}\right)\pi r_b^2 \tag{5-47}$$

Since $\pi r_b^2 = \mathfrak{U}\tau_b/q$, Eq. (5-47) can be written

$$u_y = \frac{\Delta y}{\tau_b} = \frac{\mathfrak{U}}{q|\mathbf{B}(x_0)|}\left(\frac{\partial B_z}{\partial x}\bigg|_{x_0}\right) \tag{5-48}$$

In the more general case, the drift due to gradients in the magnetic field perpendicular to the field can be expressed as

$$\mathbf{u}_G = \frac{\mathfrak{U}}{q|\mathbf{B}|}\,\mathbf{a}_B \times \nabla_\perp|\mathbf{B}| \tag{5-49}$$

where $\mathbf{a}_B$ is a unit vector in the direction of $\mathbf{B}$, $\mathbf{a}_B = \mathbf{B}/|\mathbf{B}|$, and $\nabla_\perp$ is the component of the gradient operator perpendicular to $\mathbf{a}_B$. The direction of drift is dependent on the sign of the charge and is perpendicular to both the magnetic field and the gradient of the magnetic field.

From Eq. (5-49) we can find an expression for the effective force exerted on the guiding center of the particle due to the magnetic field gradient. In our discussion of the drift velocity of guiding centers, we found that

$$\mathbf{v}_D = \frac{\mathbf{F}_\perp \times \mathbf{B}}{q|\mathbf{B}|^2} \tag{5-26}$$

If we compare Eq. (5-49) with Eq. (5-26), it is evident that the effective force caused by the gradient field is

$$\mathbf{F}_G = -\mathfrak{U}\,\nabla_\perp|\mathbf{B}| \tag{5-50}$$

Equation (5-50) is of the same form as the expression for the force on an orbiting particle in a magnetic mirror configuration [Eq. (5-34)]. The form of the force in Eq. (5-50) implies that the magnetic dipole moment of a charged particle remains constant in the gradient magnetic field.

The drift velocity of an individual charged particle of constant magnetic moment $\mathfrak{U}$ in the magnetic field configuration under consideration is given by Eq. (5-49). In a collisionless plasma there will be a distribution of particle energies and hence a distribution of magnetic moments. There will, therefore, also be a distribution of gradient drift velocities. Those particles with large magnetic moments will drift more rapidly than those particles with small magnetic moments. The average gradient drift velocity for the ions or for the electrons as a function of position coordinates and of time can be found by averaging the gradient drift velocity from Eq. (5-49) over the appropriate distribution function. The result of this manipulation is

$$\mathbf{v}_G(\mathbf{r},t) = \frac{\overline{\mathcal{U}}}{q|\mathbf{B}|}\,\mathbf{a}_B \times \nabla_\perp|\mathbf{B}| \tag{5-51}$$

where we use the velocity variable $\mathbf{v}$ to indicate that Eq. (5-51) represents the average velocity of the electrons or of the ions.

Curved Magnetic Field. $\partial B_x/\partial z$, $\partial B_y/\partial z$. The partial derivatives $\partial B_x/\partial z$ and $\partial B_y/\partial z$ cause the magnetic field lines to be curved. To show that this is the case, we shall consider the simplified example of a two-dimensional magnetic field located in the xz plane. The equation of the magnetic field lines is $d\mathbf{r} \times \mathbf{B} = 0$. For the example we are considering, the equation for the field lines is therefore

$$B_x\,dz - (B_{0z} + B_z)\,dx = 0$$

or

$$\frac{dx}{dz} = \frac{B_x}{B_z + B_{0z}} \tag{5-52}$$

Equation (5-52) can be integrated with the result that

$$x = \int \frac{B_x\,dz}{B_z + B_{0z}} + \text{constant} \tag{5-53}$$

We can expand B_x in the integrand of Eq. (5-53) in a Taylor series in z about the point where $B_x = 0$. We shall assume that this point is $z = 0$.

$$B_x = \left(\frac{\partial B_x}{\partial z}\bigg|_0\right) z \tag{5-54}$$

Only the first term in the series has been retained. The magnetic field appearing in the denominator of the integrand in Eq. (5-53) can vary

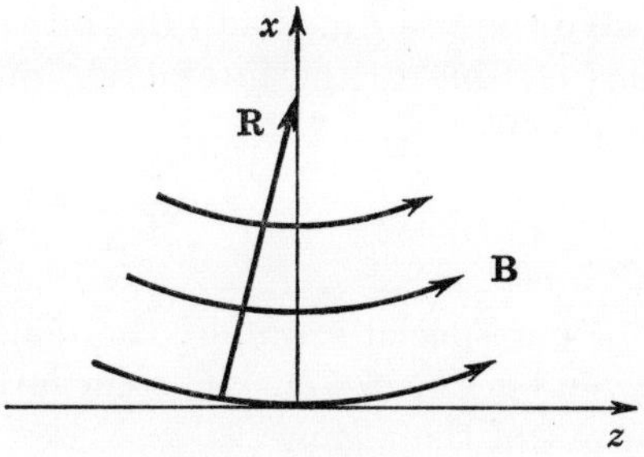

Fig. 5-8 A curved magnetic field.

only slightly with z, so that we may approximate $(B_{0z} + B_z) \cong B_{0z}$. Using this approximation and Eq. (5-54), we can integrate Eq. (5-53).

$$x = \frac{1}{B_{0z}} \left(\frac{\partial B_x}{\partial z} \bigg|_0 \right) \frac{z^2}{2} + \text{constant} \tag{5-55}$$

Equation (5-55) is the equation of the curved magnetic field lines. The curved magnetic field is shown in Fig. 5-8.

We wish now to determine the motion of a charged particle in a curved magnetic field. We shall assume that the charged particle is orbiting around the magnetic field lines with the appropriate cyclotron frequency and in addition has a speed parallel to the field lines $u_\parallel$. It is reasonable to expect that the trajectory of the guiding center of the particle will attempt to follow a curved magnetic field line as long as the curvature of that magnetic field line is not too great. Loosely speaking, there will be a centrifugal force on the particle due to the acceleration of the particle as its guiding center follows the field line.

$$\mathbf{F}_c = -\frac{mu_\parallel^2}{|\mathbf{R}|^2} \mathbf{R} \tag{5-56}$$

where **R** is the vector radius of curvature of the guiding-center trajectory and of the field line. For our example, the reciprocal of the magnitude of the radius of curvature can be found by taking the second derivative of Eq. (5-55) with respect to z.

$$\frac{1}{|\mathbf{R}|} = \frac{1}{B_{0z}} \frac{\partial B_x}{\partial z} \bigg|_0 \tag{5-57}$$

The radius of curvature is shown in Fig. 5-8.

We can use Eq. (5-26) to find the drift velocity of a charged plasma particle due to the interaction of the curvature forces and the total magnetic field.

$$\mathbf{u}_c = -\frac{mu_{\|}^2\mathbf{R}}{q|\mathbf{R}|^2} \times \frac{\mathbf{B}}{|\mathbf{B}|^2} \tag{5-58}$$

For our special example, the magnitude of the radius of curvature is given by Eq. (5-57). The curvature drift is in the positive y direction or out of the page.

Equation (5-58) is correct for the general case as long as the particle remains attached to a magnetic field line. The radius of curvature for a general magnetic field is given by

$$\frac{\mathbf{R}}{|\mathbf{R}|^2} = (\mathbf{a}_B \cdot \nabla)\mathbf{a}_B{}^* \tag{5-59}$$

Thus in the general case the curvature drift velocity is

$$\mathbf{u}_c = \frac{2\mathcal{E}_{\|}}{q|\mathbf{B}|}\,\mathbf{a}_B \times [(\mathbf{a}_B \cdot \nabla)\mathbf{a}_B] \tag{5-60}$$

We can find the average drift velocity of a collisionless group of identical plasma particles immersed in a curved magnetic field by averaging Eq. (5-60) over the appropriate particle distribution function. The average drift velocity is

$$\mathbf{v}_c(\mathbf{r},t) = \frac{2\overline{\mathcal{E}_{\|}}}{q|\mathbf{B}|}\,\mathbf{a}_B \times [(\mathbf{a}_B \cdot \nabla)\mathbf{a}_B] \tag{5-61}$$

where $\overline{\mathcal{E}_{\|}(\mathbf{r},t)}$ is the average particle energy associated with particle motion parallel to the magnetic field. Notice that we have again used the velocity variable $\mathbf{v}$ to designate the average velocity of a system of identical plasma particles.

Discussion. The magnetic dipole moment was defined for a charged particle orbiting in a static uniform magnetic field. We have used the concept of a magnetic moment in situations for which the magnetic field was not uniform. The criterion for the validity of such a procedure is that the magnetic field be approximately constant across a particle orbit. Thus a moving particle must make many orbits before changing its position in space. We see then that Figs. 5-6 to 5-8 are greatly exaggerated for the purposes of illustration.

* The vector quantity $(\mathbf{B} \cdot \nabla)\mathbf{A}$ is discussed in Appendix C.

We have made the assumptions often that the variation of the z component of the magnetic field is small and that the x and y components of the magnetic field are small compared with the z component. If these assumptions are not warranted, many of the results we have obtained will not be strictly valid. We can, however, by using the results of our analyses, get a good physical picture of the behavior of charged particles in magnetic fields which do not strictly satisfy our approximations.

We have neglected to discuss the effects of the magnetic field variation given by $\partial B_x/\partial y$, $\partial B_y/\partial x$. These terms represent a shearing force on an orbiting charged particle. The guiding center of the particle does not drift.

We have found expressions for the average velocity $\mathbf{v}(\mathbf{r},t)$ of groups of identical plasma particles in various nonuniform magnetic fields. To be more precise, the average velocities which we have calculated are the average velocities of the *particle guiding centers* contained within a small real-space volume d^3r centered about $\mathbf{r}$. This precise definition is important since it specifies which particles we are counting. Although the guiding center of a particle is within d^3r, the particle may be outside that volume; conversely, a particle may be within d^3r, while its guiding center is outside of that volume. We have defined the average velocity for a system of identical plasma particles in Chaps. 2 [Eq. (2-18)] and 4 as the average velocity of all *particles* located within the volume d^3r regardless of the position of the guiding centers of those particles. Since the two methods of calculating the average velocity differ, we should not be surprised if, in certain cases, different results are obtained by using the different averaging techniques. For a detailed discussion of the differences between the two types of averages and their mathematical effects, the reader is referred to the bibliography.

5-2-6 The Spatially Uniform Time-varying Magnetic Field. A time-varying magnetic field induces an electric field

$$\oint \mathbf{E} \cdot d\mathbf{L} = -\iint \frac{\partial \mathbf{B}}{\partial t} \cdot \mathbf{n}\, dA \qquad (4\text{-}3b)$$

A particle in a static magnetic field retains a constant energy. A particle in a time-varying magnetic field can gain or lose energy from the induced electric field. We shall show in this section that if the rate of change of the magnetic field is much slower than the cyclotron frequency of the particle, the magnetic dipole moment of the particle $\mathcal{U} = \mathcal{E}_\perp/|\mathbf{B}|$ remains approximately constant. Thus the particle

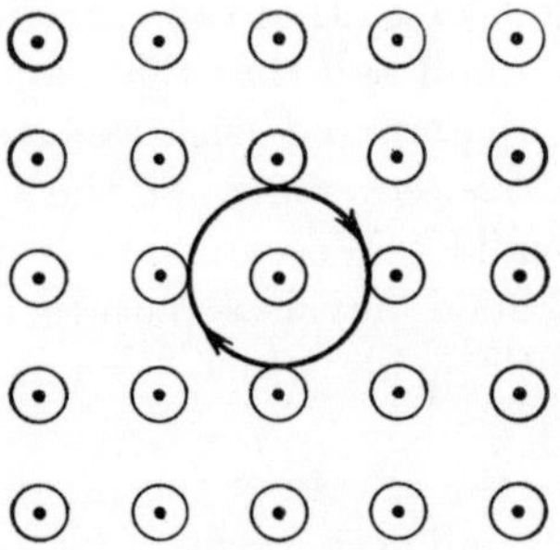

Fig. 5-9 The orbital direction of a positively charged particle.

energy must increase and decrease as the magnetic field increases and decreases.

In Fig. 5-9 the magnetic field is out of the page, and the orbit of a positively charged particle is shown. As the magnetic field is increased, an electric field is induced around the orbit in such a direction as to force the particle to increase its speed. Saying the same thing another way, Eq. (4-3*b*) indicates that the electromotive force induced around any closed path is in such a direction that the current it would produce, if it were allowed to produce a current, would generate a magnetic field which would oppose the change in the original field. Thus the induced electric field is in the direction in which the positively charged particle is orbiting. A negatively charged particle will orbit in the direction opposite to that of a positively charged particle and hence will also gain energy as the magnetic field is increased.

The energy which the particle gains in one orbit is

$$\Delta \mathcal{E}_\perp = +q \oint \mathbf{E} \cdot d\mathbf{L} \tag{5-62}$$

where $d\mathbf{L}$ is around the orbit in the direction of the particle motion. Equation (5-62) can be combined with Eq. (4-3*b*) to yield

$$\nabla \mathcal{E}_\perp = -q \iint \frac{\partial \mathbf{B}}{\partial t} \cdot \mathbf{n}\, dA \tag{5-63}$$

where dA is limited by the perimeter of the orbit. Since the particle is assumed to orbit many times before the magnetic field can change appreciably, and since the vector normal to the area is antiparallel to the magnetic field, Eq. (5-63) can be written

$$\nabla \mathcal{E}_\perp = q \frac{\partial |\mathbf{B}|}{\partial t} \iint dA = q \frac{d|\mathbf{B}|}{dt} \pi r_b^2 \tag{5-64}$$

The total derivative is used since **B** is only a function of time. The time rate of change of energy is given by $\Delta \mathcal{E}_\perp / \tau_b$. It follows that

$$\frac{d\mathcal{E}_\perp}{dt} = \frac{q\pi r_b^2}{\tau_b} \frac{d|\mathbf{B}|}{dt} \tag{5-65}$$

The factor multiplying $d|\mathbf{B}|/dt$ is the magnetic dipole moment $\mathcal{U}$, so that Eq. (5-65) becomes

$$\frac{d\mathcal{E}_\perp}{dt} = \mathcal{U} \frac{d|\mathbf{B}|}{dt} \tag{5-66}$$

The magnetic dipole moment can be written

$$\mathcal{U} = \frac{\mathcal{E}_\perp}{|\mathbf{B}|} \tag{5-10}$$

If we take the time derivative of $\mathcal{E}_\perp$ from Eq. (5-10), the result is

$$\frac{d\mathcal{E}_\perp}{dt} = \mathcal{U} \frac{d|\mathbf{B}|}{dt} + |\mathbf{B}| \frac{d\mathcal{U}}{dt} \tag{5-67}$$

A comparison of Eqs. (5-66) and (5-67) reveals that for the slowly varying magnetic field

$$\frac{d\mathcal{U}}{dt} = 0$$

The magnetic dipole moment remains constant. Thus as the magnetic field increases, the particle energy increases at the same rate. The magnetic moment and the flux through the orbit of the particle are related by a constant.

$$\varphi = \frac{2\pi m}{q^2} \mathcal{U} \tag{5-11}$$

The flux must therefore also remain constant with time. As the magnetic field increases, the orbit must shrink in size so that the same amount of flux can be included.

5-3 The Fluid Approach with Collisions

5-3-1 Assumptions and General Derivation. We shall now consider the effects of external electric and magnetic fields on a plasma whose constituent particles suffer collisions. We shall solve the momentum transfer equation (5-2) for cases in which the effective collision frequency of the particles under study is constant and the gradient of the pressure is zero. We shall allow a uniform static magnetic field in the z direction B_{0z} and a uniform electric field whose time variation goes as $e^{-i\omega t}$. Since the electric field varies as $e^{-i\omega t}$, the fluid velocity must also vary as $e^{-i\omega t}$. With the above assumptions, Eq. (5-2), written in component form, becomes

$$v_x(m\nu_c - i\omega m) = qE_x + qv_yB_{0z} \tag{5-68}$$
$$v_y(m\nu_c - i\omega m) = qE_y - qv_xB_{0z} \tag{5-69}$$
$$v_z(m\nu_c - i\omega m) = qE_z \tag{5-70}$$

The solution to Eq. (5-70) is

$$v_z = \frac{qE_z}{m(\nu_c - i\omega)} \tag{5-71}$$

Equations (5-68) and (5-69) can be solved simultaneously for v_x and for v_y.

$$v_x = \frac{q}{m}\,\frac{E_x(\nu_c - i\omega) + E_y\omega_b}{(\nu_c - i\omega)^2 + \omega_b^2} \tag{5-72}$$
$$v_y = \frac{q}{m}\,\frac{E_x(-\omega_b) + E_y(\nu_c - i\omega)}{(\nu_c - i\omega)^2 + \omega_b^2} \tag{5-73}$$

where $\omega_b = qB_{0z}/m$, the cyclotron frequency.

The information contained in Eqs. (5-71) to (5-73) can be included in one equation.[1]

$$\begin{bmatrix} v_x \\ v_y \\ v_z \end{bmatrix} = \frac{q}{m}\begin{bmatrix} \dfrac{\nu_c - i\omega}{(\nu_c - i\omega)^2 + \omega_b^2} & \dfrac{\omega_b}{(\nu_c - i\omega)^2 + \omega_b^2} & 0 \\ \dfrac{-\omega_b}{(\nu_c - i\omega)^2 + \omega_b^2} & \dfrac{\nu_c - i\omega}{(\nu_c - i\omega)^2 + \omega_b^2} & 0 \\ 0 & 0 & \dfrac{1}{\nu_c - i\omega} \end{bmatrix}\begin{bmatrix} E_x \\ E_y \\ E_z \end{bmatrix} \tag{5-74}$$

[1] For a discussion of elementary tensor operations, see Appendix D.

The current density in the plasma due to both ions and electrons is

$$\mathbf{j} = (Z_i|e|n_i\mathbf{v}_i - |e|n_e\mathbf{v}_e) \tag{4-27}$$

We can therefore write the current density as the product of a tensor and a vector.

$$\mathbf{j} = \langle\boldsymbol{\sigma}\rangle \cdot \mathbf{E} \tag{5-75}$$

where the conductivity tensor $\langle\boldsymbol{\sigma}\rangle$ is

$$\langle\boldsymbol{\sigma}\rangle = \frac{Z_i^2|e|^2n_i}{m_i}\begin{bmatrix} \dfrac{\nu_{ci} - i\omega}{(\nu_{ci} - i\omega)^2 + \omega_{bi}^2} & \dfrac{\omega_{bi}}{(\nu_{ci} - i\omega)^2 + \omega_{bi}^2} & 0 \\ \dfrac{-\omega_{bi}}{(\nu_{ci} - i\omega)^2 + \omega_{bi}^2} & \dfrac{\nu_{ci} - i\omega}{(\nu_{ci} - i\omega)^2 + \omega_{bi}^2} & 0 \\ 0 & 0 & \dfrac{1}{\nu_{ci} - i\omega} \end{bmatrix}$$
$$+ \frac{|e|^2n_e}{m_e}\begin{bmatrix} \dfrac{\nu_{ce} - i\omega}{(\nu_{ce} - i\omega)^2 + \omega_{be}^2} & \dfrac{\omega_{be}}{(\nu_{ce} - i\omega)^2 + \omega_{be}^2} & 0 \\ \dfrac{-\omega_{be}}{(\nu_{ce} - i\omega)^2 + \omega_{be}^2} & \dfrac{\nu_{ce} - i\omega}{(\nu_{ce} - i\omega)^2 + \omega_{be}^2} & 0 \\ 0 & 0 & \dfrac{1}{\nu_{ce} - i\omega} \end{bmatrix} \tag{5-76}$$

Using Eqs. (5-74) to (5-76), we shall compute the behavior of the plasma particles for a number of important situations.

5-3-2 Uniform Static Electric Field. If there is no magnetic field present, the off-diagonal terms of Eq. (5-74) are zero since the cyclotron frequency is zero. The diagonal terms are all equal, so that the particle average velocity may be written

$$\mathbf{v} = \frac{q\mathbf{E}}{m\nu_c} \tag{5-77}$$

The total current density within the plasma is

$$\mathbf{j} = \left(\frac{n_iZ_i^2|e|^2}{m_i\nu_{ci}} + \frac{n_e|e|^2}{m_e\nu_{ce}}\right)\mathbf{E} \tag{5-78}$$

Equation (5-78) is a very important expression. For plasmas in which the collision frequencies are significant, Eq. (5-78) gives the relation between the applied electric field and the current that will flow. The resistance of a plasma determines the rate at which the plasma energy can be raised by ohmic heating. In controlled-fusion machines, the energy of the plasma must be raised to thousands of electron volts. A part of this heating can be accomplished by ohmic processes.

5-3-3 Static Electric and Magnetic Fields. $\nu_c = 0$. If the collision frequency is negligible and the electric and magnetic fields are static, the average velocity of the charged particles is given by

$$\begin{bmatrix} v_x \\ v_y \\ v_z \end{bmatrix} = \frac{q}{m} \begin{bmatrix} 0 & \frac{1}{\omega_b} & 0 \\ -\frac{1}{\omega_b} & 0 & 0 \\ 0 & 0 & \infty \end{bmatrix} \begin{bmatrix} E_x \\ E_y \\ E_z \end{bmatrix} \tag{5-79}$$

or

$$v_x = \frac{qE_y}{\omega_b m} = \frac{E_y}{B_{0z}} \tag{5-80}$$

$$v_y = -\frac{qE_x}{\omega_b m} = -\frac{E_x}{B_{0z}} \tag{5-81}$$

Equations (5-80) and (5-81) can be written in vector form.

$$\mathbf{v}_\perp = \frac{\mathbf{E}_\perp \times \mathbf{B}}{|\mathbf{B}|^2} \tag{5-82}$$

This is the result we obtained previously for the average motion of a group of identical charged particles in crossed static electric and magnetic fields. The particle drift or average velocity is independent of the charge and mass of the particle, so that since the ion and electron charge densities are approximately equal in a plasma, there is no net electrical current density. There is, however, a mass flow.

Because of the assumed time variation of the velocity, we have lost all information concerning v_z. The velocity in the z direction will increase linearly with time, as we have discussed in Sec. 5-2-1.

5-3-4 Static Electric and Magnetic Fields with Collisions. We shall now consider the fluidlike motion of one group of plasma particles in crossed static electric and magnetic fields with particle collisions included. For an electric field of zero frequency, Eq. (5-74) becomes

$$\begin{bmatrix} v_x \\ v_y \\ v_z \end{bmatrix} = \frac{q}{m} \begin{bmatrix} \dfrac{\nu_c}{\nu_c^2 + \omega_b^2} & \dfrac{\omega_b}{\nu_c^2 + \omega_b^2} & 0 \\ \dfrac{-\omega_b}{\nu_c^2 + \omega_b^2} & \dfrac{\nu_c}{\nu_c^2 + \omega_b^2} & 0 \\ 0 & 0 & \dfrac{1}{\nu_c} \end{bmatrix} \begin{bmatrix} E_x \\ E_y \\ E_z \end{bmatrix} \tag{5-83}$$

Equation (5-83) can be written in component form as follows:

$$v_x = \frac{q}{m}\left(\frac{\nu_c}{\nu_c^2 + \omega_b^2} E_x + \frac{\omega_b}{\nu_c^2 + \omega_b^2} E_y\right) \tag{5-84}$$

$$v_y = \frac{q}{m}\left(\frac{-\omega_b}{\nu_c^2 + \omega_b^2} E_x + \frac{\nu_c}{\nu_c^2 + \omega_b^2} E_y\right) \tag{5-85}$$

$$v_z = \frac{q}{m}\frac{E_z}{\nu_c} \tag{5-86}$$

To get more of a physical feeling for the solution, we shall assume that $E_y = 0$. Under these conditions, Eqs. (5-84) and (5-85) become

$$v_x = \frac{q}{m}\left(\frac{\nu_c}{\nu_c^2 + \omega_b^2}\right) E_x \tag{5-87}$$

$$v_y = \frac{q}{m}\left(\frac{-\omega_b}{\nu_c^2 + \omega_b^2}\right) E_x \tag{5-88}$$

There is a velocity in the y direction due to the interaction of the electric field in the x direction and the magnetic field in the z direction. As the collision frequency becomes large compared with the cyclotron frequency,

$$v_y \rightarrow 0 \qquad v_x \rightarrow \frac{qE_x}{m\nu_c}$$

This is the result obtained in Sec. 5-3-2, for no magnetic field present. As the collision frequency becomes small compared with the cyclotron frequency,

$$v_y \rightarrow -\frac{E_x}{B_{0z}} \qquad v_x \rightarrow 0$$

This is the result obtained in Sec. 5-3-3, in the absence of collisions.

5-3-5 Polarization Drift. $\nu_c = 0$, $\omega \ll \omega_b$. If the frequency of the time-varying electric field is much less than the cyclotron frequency and if collisions can be ignored, the interaction of the time-varying electric field with the magnetic field will cause charged particles to exhibit a so-called "polarization drift." The polarization drift will be dependent on both the mass and the charge of the particles. We can write Eq. (5-74), using the assumption $\omega \ll \omega_b$, as

$$\begin{bmatrix} v_x \\ v_y \\ v_z \end{bmatrix} = \frac{q}{m} \begin{bmatrix} -\dfrac{i\omega}{\omega_b^2} & \dfrac{1}{\omega_b} & 0 \\ -\dfrac{1}{\omega_b} & -\dfrac{i\omega}{\omega_b^2} & 0 \\ 0 & 0 & -\dfrac{1}{i\omega} \end{bmatrix} \begin{bmatrix} E_x \\ E_y \\ E_z \end{bmatrix} \tag{5-89}$$

Equation (5-89), in component form with $\omega_b = qB_{0z}/m$, becomes

$$v_x = \frac{-i\omega m E_x}{qB_{oz}^2} + \frac{E_y}{B_{0z}} \tag{5-90}$$

$$v_y = \frac{-i\omega m E_y}{qB_{oz}^2} - \frac{E_x}{B_{0z}} \tag{5-91}$$

$$v_z = -\frac{qE_z}{i\omega m} \tag{5-92}$$

Equations (5-90) and (5-91) can be combined into one vector equation.

$$\mathbf{v}_\perp = \frac{-i\omega m \mathbf{E}_\perp}{q|\mathbf{B}|^2} + \frac{\mathbf{E}_\perp \times \mathbf{B}}{|\mathbf{B}|^2} \tag{5-93}$$

The second term on the right of Eq. (5-93) is the usual charge and mass independent drift in crossed electric and magnetic fields. The additional term on the right of Eq. (5-93) is called the "polarization drift velocity." Since $\partial \mathbf{E}_\perp/\partial t = -i\omega \mathbf{E}_\perp$, the polarization drift velocity can be rewritten

$$\mathbf{v}_P = \frac{m}{q|\mathbf{B}|^2} \frac{\partial \mathbf{E}_\perp}{\partial t} \tag{5-94}$$

Notice that the polarization drift is in a direction perpendicular to the static magnetic field. The polarization drift is proportional to the time rate of change or frequency of the applied electric field. Since oppositely charged particles drift in opposite directions, the polarization drift will cause a net current flow in the plasma. Since the frequency variation of the electric field is assumed to be much smaller than the cyclotron frequency, a charged particle will orbit the magnetic field with an approximately constant dipole moment as it drifts in the direction indicated by Eq. (5-93). The orbiting motion, of course, cannot appear in the solution for the average or fluid velocity.

PROBLEMS

5-1 In Sec. 4-3-4 we discussed the fact that, from the hydrodynamic point of view, nine equations are necessary to specify completely the behavior of a three-component plasma. In Chap. 5 we solved for $\mathbf{v}_e$ and $\mathbf{v}_i$ by using only two of these nine equations. List all assumptions and approximations which make possible such a solution.

5-2 We made the approximation in Chap. 5 that ν_c is not a function of $\mathbf{v}$. This approximation is often a very good one if $\bar{u} \gg v$. Discuss.

5-3 Solve Eq. (5-6) in rectangular coordinates.

5-4 Find numerical values of ω_b for electrons and for oxygen ions (O^+) if (*a*) $|\mathbf{B}| = 1$ weber/m^2 (a very large magnetic field) and (*b*) $|\mathbf{B}| = 10^{-4}$ weber/m^2 (a magnetic field which can be created in the laboratory without special equipment).

5-5 We have found that the magnitude of the drift velocity of a charged particle in perpendicular uniform electrostatic and magnetostatic fields is $v_E = E/|\mathbf{B}|$. In deriving this result, we have neglected to take account of the fact that the mass of a particle increases as the velocity of the particle approaches the speed of light $c = 3 \times 10^8$ m/sec.

$$m = \frac{m_0}{[1 - (v/c)^2]^{1/2}}$$

where m_0 is the mass of the stationary particle. Could this relativistic mass increase be important in a laboratory plasma? Do not forget that the radius of the cyclotron orbit must fit within the laboratory.

5-6 We have derived Eq. (5-26) for the case of static uniform external force applied in conjunction with the uniform magnetostatic field. Is Eq. (5-26) valid for external forces which are not spatially constant?

5-7 Assume that a collisionless cylindrically shaped plasma is confined by a uniform static magnetic field as shown in Fig. P5-7. The plasma particles orbit around the magnetic field lines and thus cannot drift perpendicular to the magnetic field lines. At some time $t = t_0$, a uniform gravitation field is applied to the system of magnetic field and plasma in a direction perpendicular to the magnetic field. The plasma particles will begin to drift in the combined gravitational and magnetic fields. The charge separation which results from this drift produces an electric field in the plasma. The plasma particles will drift in the combined electric and magnetic fields. Show qualitatively that if the applied gravitational field is vertically downward,

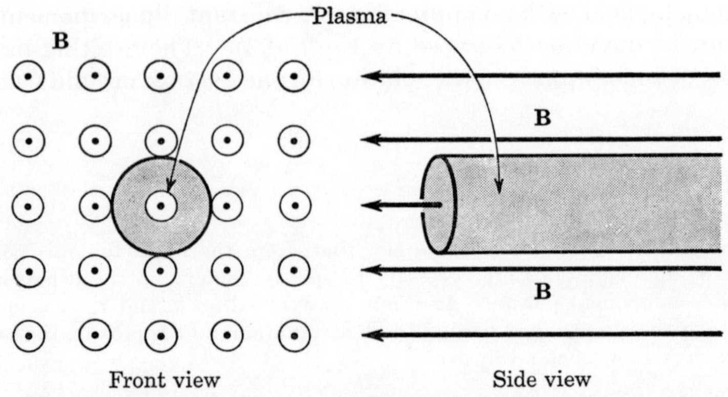

Fig. P5-7 A cylindrically shaped plasma confined by a magnetic field.

all the plasma particles will move vertically downward across the magnetic field lines.

5-8 Show qualitatively that it is impossible to contain a plasma in the magnetic field of a toroid.

$$\mathbf{B} = \frac{k}{r}\,\mathbf{a}_\theta$$

where k is a constant. The magnetic field is shown in Fig. P5-8.

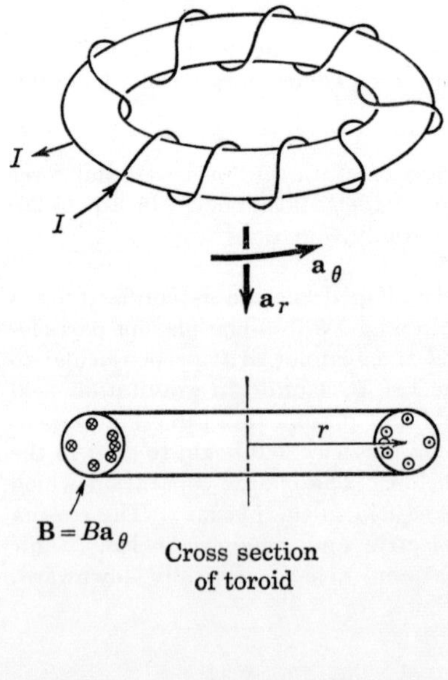

Fig. P5-8 A toroid and its magnetic field.

5-9 Derive the polarization drift velocity using the individual-particle approach. Show that the magnetic moment remains invariant to first approximation.

BIBLIOGRAPHY

Allis, W. P.: Motion of Ions and Electrons, "Handbuch der Physik," vol. 21 (in English), Springer-Verlag OHG, Berlin, 1956.
chap. I, p. 384—Electron and Ion Orbits.
chap. II, p. 392—The Langevin Equation.

Chandrasekhar, S.: "Plasma Physics," The University of Chicago Press, Chicago, 1960.
chap. 2—The First Order Orbit Theory. On the material in Sec. 5-2.
chap. 2, sec. 8—The Magnetic Moment and the Magnetization Current.

Harmon, W. W.: "Fundamentals of Electronic Motion," McGraw-Hill Book Company, Inc., New York, 1953.

Longmire, C. L.: "Elementary Plasma Physics," Interscience Publishers, Inc., New York, 1963.
chap. 2—First-order Orbit Theory. On the material in Sec. 5-2.

Plonsey, R., and R. E. Collin: "Principles and Applications of Electromagnetic Fields," McGraw-Hill Book Company, Inc., New York, 1961.
sec. 7.1—Magnetic Dipole Polarization per Unit Volume.

Spitzer, L., Jr.: "Physics of Fully Ionized Gases," 2d ed., Interscience Publishers, Inc., New York, 1962.
chap. 1—Motion of a Particle.
chap. 2, sec. 3—A discussion of the differences between the average velocity of the particles and the average velocity of the guiding centers of the particles.

chapter

6

Orbit Theory

6-1 Introduction

In many cases, essentially collisionless plasmas exist in regions containing the types of nonuniform magnetic fields discussed in Chap. 5. Stars and galaxies produce relatively strong directional magnetic fields in which galactic plasmas move. The controlled-fusion program makes use of approximately uniform magnetic fields to confine the high-energy plasma. We have shown in Chap. 5 that under certain conditions the magnetic dipole moment of a charged particle moving in electric and magnetic fields remains constant. Under these conditions the motion of the particle can be described by specifying the motion of the guiding center of the particle. Orbit theory is that branch of plasma physics which concerns itself with situations in which the magnetic dipole moments of the plasma particles remain constant. Orbit theory concerns itself with the behavior of fixed magnetic dipole moments in various field configurations. Once it has been proved that the magnetic dipole moments of the plasma particles are constant for a given field configuration, the current density and the charge density due to the motion of the guiding centers of the orbiting particles can be calculated. This calculation requires a knowledge of the density of the particles and of the average drift velocities of the guiding centers of the particles in the particular field configuration under study. A knowledge of the current density and the charge density in terms of the fields is equivalent to a knowledge of the solution of the hydrodynamic equations. The current and charge densities can be combined with Maxwell's equations to provide a set of equations which can be solved for the behavior of the electromagnetic fields.

In this chapter we shall discuss the principles and validity of orbit theory. We shall then summarize the various guiding-center drifts and current densities arising in field configurations for which orbit theory is valid. Following this summary, two examples of the application of orbit theory will be given. We shall defer until Chap. 13 the calculation of the behavior of hydromagnetic plasma waves from the viewpoint of orbit theory.

6-2 The Principles and Validity of Orbit Theory

In Sec. 5-2-5 we examined the behavior of charged particles in the following magnetic field configurations: a static magnetic field in the z direction with a gradient of the field perpendicular to the z direction, a static magnetic field with curvature, and a magnetic mirror field. In each case we assumed that the total magnetic field was composed of a strong static magnetic field in the z direction B_{0z} and the appropriate perturbational fields B_x, B_y, B_z. In calculating the particle motion in the "gradient" field, we assumed that the only important field variation was $\partial B_z/\partial x$, $\partial B_z/\partial y$; in calculating the particle motion in the curved field, we assumed that the only important field variation was $\partial B_x/\partial z$, $\partial B_y/\partial z$; in calculating the particle motion in the mirror field, we assumed that the only important field variation was $\partial B_x/\partial x$, $\partial B_y/\partial y$, $\partial B_z/\partial z$. In each case we showed that the magnetic moment of the orbiting particle remained constant.

In general, an arbitrary magnetic field of the form B_{0z}, B_x, B_y, B_z may concurrently exhibit finite values for

$$\frac{\partial B_z}{\partial x}, \frac{\partial B_z}{\partial y}, \frac{\partial B_x}{\partial z}, \frac{\partial B_y}{\partial z}, \frac{\partial B_x}{\partial x}, \frac{\partial B_y}{\partial y}, \frac{\partial B_z}{\partial z}$$

(We have excluded $\partial B_y/\partial x$, $\partial B_x/\partial y$ from the discussion since these two terms do not lead to guiding-center motion.) It is not obvious that the results of our previous analyses can be applied to this general situation. We shall show that they can be.

According to the rules of orbit theory, the total velocity of a charged particle in an arbitrary static magnetic field of the form B_{0z}, B_x, B_y, B_z can be decomposed into four parts: an orbital motion whose associated magnetic moment remains constant, a perturbational motion due to the "gradient" type of variation in the magnetic field, a perturbational motion due to the "curvature" type of variation in the magnetic field, and a perturbational motion due to the "mirror" type of variation in the magnetic field. The latter three motions are perturbational with respect to the orbital velocity because the magnetic fields which cause these motions are small with respect to B_{0z} and because these small fields vary only slightly over the area of the orbital motion.

Each of the perturbational velocities can be considered independent of the other, and thus each can be calculated from the expressions derived in Sec. 5-2-5.

We shall now justify the "rules" of orbit theory stated in the preceding paragraph. If the magnetic moment of a charged particle remains constant in the presence of each of the three types of perturbational fields, it is reasonable to expect that the magnetic moment of the particle will remain constant in combinations of the perturbational fields. It is a little more difficult to see that the total perturbational velocity of the guiding center of a charged particle in an arbitrary magnetic field can be decomposed into three distinct velocities, each of the velocities being associated with a particular property of the magnetic field. We can best show that this is so mathematically. We shall assume that we have solved for the velocity of the guiding center of a particle in an arbitrary magnetic field at one particular time. We can write this velocity as

$$\mathbf{u}\left(B_{0z}, \mathfrak{U}, \frac{\partial B_z}{\partial x}, \frac{\partial B_z}{\partial y}, \frac{\partial B_x}{\partial z}, \frac{\partial B_y}{\partial z}, \frac{\partial B_x}{\partial x}, \frac{\partial B_y}{\partial y}, \frac{\partial B_z}{\partial z}, t\right)$$

where B_{0z}, $\mathfrak{U}$, and t are constants. The velocity of the guiding center can be expanded in a power series in the seven quantities which affect the velocity. Since the perturbational fields are assumed small and their spatial variation is assumed small, the second-order terms and all higher-order terms in the power-series expansion can be ignored. The second- and higher-order terms contain products of the various different types of field variation. Thus the velocity of the guiding center of the particle can be written as a linear sum of velocity terms, each of which depends on the variation of one component of the magnetic field in one direction. These separate terms can be collected into three sets of terms, each set of terms being associated with one of the three properties of the magnetic field.

We have considered only static nonuniform magnetic fields thus far in our examination of the principles of orbit theory. We can, however, extend our previous results to include the case of a time-varying nonuniform magnetic field if the frequency of the time-varying magnetic field is very slow compared with the cyclotron frequency of the particles orbiting in that field. The justification for this procedure is the realization that a slowly time-varying magnetic field can neither cause the guiding center of a particle to exhibit new types of motion nor change the value of the magnetic moment of the particle. If, for example, the perturbational fields B_x, B_y, B_z grow smaller with time while B_{0z} remains constant, the motion of the guiding centers of the particles due to the spatial variation of B_x, B_y, B_z will merely grow smaller with time. No additional effects will occur as long as the magnetic fields change slowly compared with the cyclotron frequency of the plasma particles. The guiding-center motions can therefore be calculated from the expressions derived in Sec. 5-2-5, even though the magnetic fields are functions of time.

In the arbitrary nonuniform magnetic fields discussed above, the magnetic moment of an orbiting charged particle remains constant as the guiding center of the particle orbit moves because of the effects of the perturbational fields. We have shown in Chap. 5 that certain configurations of external force fields acting in conjunction with a static magnetic field will also cause the guiding center of the particle to move while the magnetic moment of the particle orbit remains constant. These guiding-center drifts are not perturbational unless the external force fields are perturbational. If the external forces are perturbational, the total velocity of the guiding center of the constant magnetic moment can be found by adding vectorially the velocities resulting from the interaction of the force fields with B_{0z} and the velocities due to the nonuniform magnetic field.

We can now summarize our discussion of orbit theory as follows. A charged particle moving in the following field configurations retains a constant magnetic moment:

1. Static magnetic field with weak gradient perpendicular to strong B_{0z}

2. Static magnetic field with weak curvature, strong B_{0z}

3. Uniform static magnetic field with perturbational perpendicular static external force

4. Uniform static magnetic field with perturbational perpendicular time-varying electric field, $\omega \ll \omega_b$

5. Magnetic mirror configuration, strong B_{0z}

6. Uniform time-varying magnetic field, $\omega \ll \omega_b$

The fact that two or more of the above configurations are present concurrently in no way affects the validity of the mathematical expressions that we have derived for the particle motions in Chap. 5. Further, the slow time variation ($\omega \ll \omega_b$) of the nonuniform magnetic field will neither cause new guiding-center motions to occur nor change the magnetic moments of the plasma particles.

Notice that there are two general criteria which all of the field configurations that we are discussing must satisfy:

1. The magnetic field must be essentially constant over the area of any one orbit.

2. The frequency of the orbital motion of any particle must be much greater than the frequency of any magnetic field or electric field variation.

These criteria must be satisfied if the magnetic moment of the orbiting particle is to remain constant.

6-3 Drift Velocities and Current Densities

We shall now list the guiding-center average drift velocities for those field configurations in which the magnetic dipole moments of the charged plasma particles can be considered constant:

1. Gradient drift:

$$\mathbf{v}_G = \frac{\overline{\mathcal{U}}}{q|\mathbf{B}|}\,\mathbf{a}_B \times \nabla_\perp |\mathbf{B}| \tag{5-51}$$

2. Curvature drift:

$$\mathbf{v}_c = \frac{2\overline{\mathcal{E}_\parallel}}{q|\mathbf{B}|}\,\mathbf{a}_B \times [(\mathbf{a}_B \cdot \nabla)\mathbf{a}_B] \tag{5-61}$$

3. External field drift:

$$\mathbf{v}_D = \frac{\mathbf{F}_\perp \times \mathbf{B}}{q|\mathbf{B}|^2} \tag{5-26}$$

4. Polarization drift:

$$\mathbf{v}_P = \frac{m}{q|\mathbf{B}|^2}\frac{\partial \mathbf{E}_\perp}{\partial t} \tag{5-94}$$

A unique drift velocity cannot be found for the particles in a magnetic mirror geometry since the particles are always undergoing acceleration.

$$\overline{F_z} = -\overline{\mathcal{U}}\,\frac{dB_z}{dz} \tag{5-42}$$

The time-varying magnetic field does not cause a particle drift.

In a plasma the current density is given by

$$\mathbf{j} = Z_i|e|n_i\mathbf{v}_i - |e|n_e\mathbf{v}_e \tag{4-27}$$

We shall use Eq. (4-27) to calculate the current densities caused by the various guiding-center drifts:

1. Gradient-drift current density:

$$\mathbf{j}_G = \left(\frac{n_i\overline{\mathfrak{U}_i}}{|\mathbf{B}|} + \frac{n_e\overline{\mathfrak{U}_e}}{|\mathbf{B}|}\right)\mathbf{a}_B \times \nabla_\perp|\mathbf{B}| \tag{6-1}$$

2. Curvature-drift current density:

$$\mathbf{j}_c = \frac{2(\overline{\mathcal{E}_{i\|}}n_i + \overline{\mathcal{E}_{e\|}}n_e)}{|\mathbf{B}|}\mathbf{a}_B \times [(\mathbf{a}_B \cdot \nabla)\mathbf{a}_B] \tag{6-2}$$

3. External-field–drift current density:

$$\mathbf{j}_D = \frac{(n_i\mathbf{F}_{i\perp} + n_e\mathbf{F}_{e\perp})}{|\mathbf{B}|^2} \times \mathbf{B} \tag{6-3}$$

4. Polarization-drift current density:

$$\mathbf{j}_P = \left(\frac{n_im_i + n_em_e}{|\mathbf{B}|^2}\right)\frac{\partial\mathbf{E}_\perp}{\partial t} \tag{6-4}$$

In addition to these current densities, there is the effective current density caused by a nonuniform vector magnetic moment per unit volume **M** in the plasma.

$$\mathbf{j}_M = \nabla \times \mathbf{M} \tag{5-13}$$

where

$$\mathbf{M} = -n_i\overline{\mathfrak{U}_i}\frac{\mathbf{B}}{|\mathbf{B}|} - n_e\overline{\mathfrak{U}_e}\frac{\mathbf{B}}{|\mathbf{B}|} \tag{5-12}$$

The magnetization current density is therefore given by

$$\mathbf{j}_M = -\nabla \times \frac{n_i\overline{\mathfrak{U}_i}\mathbf{B}}{|\mathbf{B}|} - \nabla \times \frac{n_e\overline{\mathfrak{U}_e}\mathbf{B}}{|\mathbf{B}|} \tag{6-5}$$

6-4 A Simple Stationary State

In this section we shall use the techniques of orbit theory to derive a relationship between the kinetic energy density of a group of plasma particles orbiting in a static magnetic field and the energy density associated with the static magnetic field in which the particles orbit.

For simplicity we shall assume that the magnetic field under consideration is only in the z direction and that it varies only as a function of x and y coordinates. The current densities which are present due to the nonuniform field are the gradient current density and the magnetization current density. The total current density within the plasma can therefore be found by adding Eqs. (6-1) and (6-5).

$$\mathbf{j} = \mathbf{j}_G + \mathbf{j}_M = \left(\frac{n_i\overline{\mathcal{U}_i}}{|\mathbf{B}|} + \frac{n_e\overline{\mathcal{U}_e}}{|\mathbf{B}|}\right)\mathbf{a}_z \times \nabla_\perp|\mathbf{B}| - \nabla \times \frac{n_i\overline{\mathcal{U}_i}\mathbf{B}}{|\mathbf{B}|} - \nabla \times \frac{n_e\overline{\mathcal{U}_e}\mathbf{B}}{|\mathbf{B}|} \tag{6-6}$$

where $\mathbf{a}_z$ is a unit vector in the z direction. Since $\overline{\mathcal{U}} = \overline{\mathcal{E}_\perp}/|\mathbf{B}|$, Eq. (6-6) can be rewritten

$$\mathbf{j} = \left(\frac{n_i\overline{\mathcal{E}_{i\perp}} + n_e\overline{\mathcal{E}_{e\perp}}}{|\mathbf{B}|^2}\right)\mathbf{a}_z \times \nabla_\perp|\mathbf{B}| - \nabla \times \left(\frac{n_i\overline{\mathcal{E}_{i\perp}} + n_e\overline{\mathcal{E}_{e\perp}}}{|\mathbf{B}|}\right)\mathbf{a}_z \tag{6-7}$$

The second term on the right of Eq. (6-7) can be written

$$\nabla \times \left(\frac{n_i\overline{\mathcal{E}_{i\perp}} + n_e\overline{\mathcal{E}_{e\perp}}}{|\mathbf{B}|}\right)\mathbf{a}_z = -\mathbf{a}_z \times \nabla_\perp \left(\frac{n_i\overline{\mathcal{E}_{i\perp}} + n_e\overline{\mathcal{E}_{e\perp}}}{|\mathbf{B}|}\right) \tag{6-8}$$

The proof of this vector identity is left for the problems. If we set $W_\perp = (n_i\overline{\mathcal{E}_{i\perp}} + n_e\overline{\mathcal{E}_{e\perp}})$, where $W_\perp$ is the kinetic energy density associated with the motion of the plasma particles perpendicular to the magnetic field, we can write Eq. (6-7) as

$$\mathbf{j} = \left(\frac{W_\perp}{|\mathbf{B}|^2}\right)\mathbf{a}_z \times \nabla_\perp|\mathbf{B}| + \mathbf{a}_z \times \nabla_\perp\left(\frac{W_\perp}{|\mathbf{B}|}\right) \tag{6-9}$$

The second term on the right of Eq. (6-9) can be expanded as follows:

$$\mathbf{a}_z \times \nabla_\perp\left(\frac{W_\perp}{|\mathbf{B}|}\right) = \frac{\mathbf{a}_z}{|\mathbf{B}|} \times \nabla_\perp W_\perp - \left(\frac{W_\perp}{|\mathbf{B}|^2}\right)\mathbf{a}_z \times \nabla_\perp|\mathbf{B}| \tag{6-10}$$

The combination of Eqs. (6-9) and (6-10) yields

$$\mathbf{j} = \frac{\mathbf{a}_z}{|\mathbf{B}|} \times \nabla_\perp W_\perp \tag{6-11}$$

The behavior of a static magnetic field in the presence of a current density is specified by Eq. (4-2a) with the time-varying field set to zero.

$$\nabla \times \mathbf{H} = \mathbf{j} \tag{6-12}$$

In the present case, we can write

$$\nabla \times \mathbf{H} = -\mathbf{a}_z \times \nabla_\perp |\mathbf{H}| \tag{6-13}$$

so that Eq. (6-12) becomes

$$-\mathbf{a}_z \times \nabla_\perp |\mathbf{H}| = \mathbf{j} \tag{6-14}$$

Equation (6-11) can be combined with Eq. (6-14). The resulting equation is

$$\mathbf{a}_z \times [|\mathbf{B}|\nabla_\perp |\mathbf{H}| + \nabla_\perp W_\perp] = 0 \tag{6-15}$$

If we let $\mathbf{B} = \mu_0 \mathbf{H}$, then

$$|\mathbf{B}|\nabla_\perp |\mathbf{H}| = \nabla_\perp \left(\frac{|\mathbf{B}|^2}{2\mu_0} \right) \tag{6-16}$$

and

$$\mathbf{a}_z \times \nabla_\perp \left(\frac{|\mathbf{B}|^2}{2\mu_0} + W_\perp \right) = 0 \tag{6-17}$$

The solution to Eq. (6-17) is

$$\left(\frac{|\mathbf{B}|^2}{2\mu_0} + W_\perp \right) = \text{constant} \tag{6-18}$$

Thus at places where the kinetic energy density of the particles is the greatest, $|\mathbf{B}|^2/2\mu_0$, the energy density of the magnetic field is the least. From a physical point of view, the plasma is a diamagnetic medium whose diamagnetic properties depend on the density and the orbital kinetic energy of the particles. Notice that the dimensions of energy density are also the dimensions of pressure. We shall have more to say about the physical significance of Eq. (6-18) and the interpretation of $W_\perp$ and $|\mathbf{B}|^2/2\mu_0$ as pressures in Chaps. 8 and 13.

6-5 A Simple Time-varying State

In this section we shall use the techniques of orbit theory to derive a relationship between the particle density of a system of identical plasma particles interacting with a nonuniform slowly time-varying magnetic field and the magnetic field with which the particles interact. We shall assume that the magnetic field under consideration is only in the z direction, that it varies only as a function of the y coordinate and of time, and that it is of the form $|\mathbf{B}| = B_{0z} + B_z(y,t)$. If there is a time-varying magnetic field present within the plasma, there must also be present a time-varying electric field. We shall allow an electric field in the x direction which is only a function of the y coordinate and of time: $|\mathbf{E}| = E_x(y,t)$. The field configuration which we choose to consider is consistent with Eq. (4-3a).

$$\nabla \times \mathbf{E} = -\frac{\partial \mathbf{B}}{\partial t} \tag{4-3a}$$

In order to investigate the density variation of a system of identical plasma particles, we must make use of the continuity equation for that system of particles

$$\frac{\partial n(\mathbf{r},t)}{\partial t} + \nabla \cdot [n(\mathbf{r},t)\mathbf{v}(\mathbf{r},t)] = 0 \tag{6-19}$$

Equation (6-19) is applicable to the plasma electrons $n = n_e$, $\mathbf{v} = \mathbf{v}_e$ or to the plasma ions $n = n_i$, $\mathbf{v} = \mathbf{v}_i$.

The various guiding-center average drift velocities are listed in Sec. 6-3. For the field configuration we are considering, the gradient drift is

$$\mathbf{v}_G(y,t) = -\frac{\overline{\mathcal{U}}}{q|\mathbf{B}(y,t)|}\frac{\partial B_z(y,t)}{\partial y}\mathbf{a}_x \tag{6-20}$$

where $\mathbf{a}_x$ is a unit vector in the x direction. The curvature drift is zero,

$$\mathbf{v}_c = 0 \tag{6-21}$$

since the magnetic field lines are assumed straight. The external field drift, with $\mathbf{F}_\perp = q\mathbf{E}$, is

$$\mathbf{v}_E = \frac{\mathbf{E} \times \mathbf{B}}{|\mathbf{B}|^2} = -\frac{E_x(y,t)}{B_{0z} + B_z(y,t)}\mathbf{a}_y \tag{6-22}$$

where $\mathbf{a}_y$ is a unit vector in the y direction. The polarization drift is

$$\mathbf{v}_P = \frac{m}{q|\mathbf{B}(y,t)|}\frac{\partial E_x(y,t)}{\partial t}\mathbf{a}_x \tag{6-23}$$

We shall assume that the average magnetic moment is independent of position coordinates. Since the magnetic moment of each individual plasma particle remains constant, this is a reasonable assumption.

Both the gradient drift and the polarization drift are in the x direction, but neither is a function of the x coordinate. Since the drift velocities in the x direction are independent of the x coordinate, the particle density cannot be a function of the x coordinate. It follows that

$$\nabla \cdot [n(\mathbf{v}_G + \mathbf{v}_P)] = 0 \tag{6-24}$$

The only drift velocity which can lead to a divergence of the particle current density is the external field drift. Equation (6-19) can therefore be written

$$\frac{\partial n}{\partial t} + \nabla \cdot (n\mathbf{v}_E) = 0 \tag{6-25}$$

We shall now focus our attention on Eq. (4-3*a*), the equation which relates the time variation of the magnetic field to the spatial variation of the electric field. We wish to express **E** in Eq. (4-3*a*) in terms of **B** and $\mathbf{v}_E$. This we can do by noting that since

$$\mathbf{v}_E = \frac{\mathbf{E} \times \mathbf{B}}{|\mathbf{B}|^2} \tag{6-22}$$

it follows from crossing each side of Eq. (6-22) into **B** that

$$\mathbf{E} = -\mathbf{v}_E \times \mathbf{B} \tag{6-26}$$

If Eq. (6-26) is substituted into Eq. (4-3*a*), the resultant relation is

$$\nabla \times (\mathbf{v}_E \times \mathbf{B}) = \frac{\partial \mathbf{B}}{\partial t} \tag{6-27}$$

The left-hand side of Eq. (6-27) can be expanded as follows:

$$\nabla \times (\mathbf{v}_E \times \mathbf{B}) = (\mathbf{B} \cdot \nabla)\mathbf{v}_E + \mathbf{v}_E(\nabla \cdot \mathbf{B}) - (\mathbf{v}_E \cdot \nabla)\mathbf{B} - \mathbf{B}(\nabla \cdot \mathbf{v}_E) \tag{6-28}$$

The first two terms on the right of Eq. (6-28) are zero for the fields under consideration since

$$(\mathbf{B} \cdot \nabla)\mathbf{v}_E = (B_{0z} + B_z)\frac{\partial \mathbf{v}_E}{\partial z} = 0$$

and

$$\nabla \cdot \mathbf{B} = 0$$

Equation (6-28) can be written

$$\nabla \times (\mathbf{v}_E \times \mathbf{B}) = -(\mathbf{v}_E \cdot \nabla)\mathbf{B} - \mathbf{B}(\nabla \cdot \mathbf{v}_E) = -\frac{\mathbf{B}}{|\mathbf{B}|}\nabla \cdot (\mathbf{v}_E|\mathbf{B}|) \quad (6\text{-}29)$$

since **B** is only in the z direction. Equations (6-29) and (6-27) can be combined to yield

$$\frac{\partial|\mathbf{B}|}{\partial t} + \nabla \cdot (\mathbf{v}_E|\mathbf{B}|) = 0 \quad (6\text{-}30)$$

Equation (6-30) is a continuity equation for $|\mathbf{B}|$. Since Eqs. (6-30) and (6-25) are of exactly the same form, it is apparent that the behavior of the particle density and of the magnetic field must be identical. We shall now prove mathematically that this is so.

The second term in Eq. (6-30) can be expanded with the result that

$$\frac{\partial|\mathbf{B}|}{\partial t} + \mathbf{v}_E \cdot \nabla|\mathbf{B}| + |\mathbf{B}|\nabla \cdot \mathbf{v}_E = 0 \quad (6\text{-}31)$$

The total derivative of $|\mathbf{B}|$ with respect to time is

$$\frac{d|\mathbf{B}|}{dt} = \frac{\partial|\mathbf{B}|}{\partial t} + \frac{\partial|\mathbf{B}|}{\partial y}\frac{dy}{dt} = \frac{\partial|\mathbf{B}|}{\partial t} + \mathbf{v}_E \cdot \nabla|\mathbf{B}| \quad (6\text{-}32)$$

The substitution of Eq. (6-32) into Eq. (6-31) yields

$$\frac{1}{|\mathbf{B}|}\frac{d|\mathbf{B}|}{dt} = -\nabla \cdot \mathbf{v}_E \quad (6\text{-}33)$$

The continuity equation for the system of particles under consideration [Eq. (6-25)] can similarly be written

$$\frac{1}{n}\frac{dn}{dt} = -\nabla \cdot \mathbf{v}_E \quad (6\text{-}34)$$

Equations (6-33) and (6-34) can be combined to yield

$$\frac{1}{n}\frac{dn}{dt} - \frac{1}{|\mathbf{B}|}\frac{d|\mathbf{B}|}{dt} = 0 \quad (6\text{-}35)$$

Equation (6-35) multiplied by $n/|\mathbf{B}|$ is

$$\frac{1}{|\mathbf{B}|}\frac{dn}{dt} - \frac{n}{|\mathbf{B}|^2}\frac{d|\mathbf{B}|}{dt} = 0 \tag{6-36}$$

or

$$\frac{d}{dt}\left(\frac{n}{|\mathbf{B}|}\right) = 0 \tag{6-37}$$

Equation (6-37) is the relationship between the particle density and the nonuniform slowly time-varying magnetic field.

We shall now examine the physical consequences of Eq. (6-37). Let us consider a fixed number of identical plasma particles which at $t = 0$ are contained within a volume d^3r. As those particles originally contained with d^3r move through space, the quotient of their density and the magnetic field in which they orbit remains constant. Saying the same thing in another way, the particles are attached to the magnetic field lines. If the magnetic field lines become dense (the field strong), the particle density will increase proportionately. If the magnetic field lines become sparse (the field weak), the particle density will become sparse proportionately. We shall have more to say about this type of particle-field interaction in Chap. 13.

6-6 Discussion

When using Maxwell's equations, it is necessary to specify the relation between **B** and **H** and between **D** and **E**. In Sec. 6-4 we have assumed that $\mathbf{B} = \mu_0\mathbf{H}$, that the permeability of the plasma is equal to the permeability of free space. If the ions and the neutral particles of the plasma have magnetic properties, this is not the case. If the ions and the neutral particles have magnetic properties, the real permeability of the plasma will be determined by those properties. The real permeability of the plasma will be a function of position coordinates since the particle densities are functions of position. From a practical point of view, the permeability of almost all plasmas is essentially equal to the permeability of free space since the particle densities characteristic of the plasma state are relatively low.

In Secs. 6-4 and 6-5 we discussed examples of the applications of orbit theory. In general, the techniques of orbit theory can be used to determine the properties of the interaction between charged plasma particles and any field configuration of the form B_{0z}, B_x, B_y, B_z, E_x, E_y as long as $|\mathbf{B}(\mathbf{r},t)| \cong B_{0z}$, $\mathbf{E}$ is small, and the fields vary slowly in time and in space. The procedure to be employed is as follows: The five current densities listed in Eqs. (6-1) to (6-5) are combined with Eq. (4-2*a*)

$$\nabla \times \mathbf{H} = \mathbf{j} + \frac{\partial \mathbf{D}}{\partial t} \tag{4-2a}$$

so as to eliminate the current density from Eq. (4-2*a*). The resulting equation and Eq. (4-3*a*)

$$\nabla \times \mathbf{E} = -\frac{\partial \mathbf{B}}{\partial t} \tag{4-3a}$$

provide a set of two equations whose unknowns are the electric field, the magnetic field, and the particle densities (we assume that the average magnetic moments are known). If the particle densities are constant, we have derived a set of two equations from which the electric and magnetic fields can be determined. If the particle densities are not constant, the continuity equations must be considered in order to effect a solution. Once the properties of the fields are determined, the behavior of the plasma particles can then be calculated. In Chap. 13 we shall determine the behavior of hydromagnetic plasma waves using the techniques outlined above.

PROBLEMS

6-1 Prove the vector identity

$$\nabla \times (A\mathbf{a}_z) = -\mathbf{a}_z \times \nabla_\perp A$$

where $\mathbf{A}$ is a vector in the z direction, $\mathbf{A} = A\mathbf{a}_z$.

6-2 Find the relationship between the kinetic energy density of a group of plasma particles orbiting in a curved magnetic field and the energy density associated with the magnetic field. Assume that the magnetic field is in the θ direction (cylindrical coordinate system) and is only a function of r. In assuming that the magnetic field is only a function of r, we are implicitly assuming that the plasma is infinite in the z direction. If the plasma has a finite dimension in the z direction, we must assume that the magnetic field and hence the particle densities are functions of both r and z. Under these circumstances, a steady-state solution for the plasma behavior cannot be found (see Prob. 5-8).

ANSWER: $$\frac{1}{\mu_0 r}\frac{d}{dr}(r|\mathbf{B}|) + \frac{1}{r|\mathbf{B}|}\frac{d}{dr}(rW_\perp) - \frac{2W_\parallel}{r|\mathbf{B}|} = 0$$

or

$$\frac{d}{dr}\left[r\left(\frac{|\mathbf{B}|^2}{2\mu_0} + W_\perp\right)\right] + \left(\frac{|\mathbf{B}|^2}{2\mu_0} - 2W_\parallel\right) = 0$$

6-3 How are the results of Sec. 6-4 affected if the permeability of the plasma is proportional to the particle densities?

6-4 Show that Eq. (6-22) is valid (see Prob. 5-6 and Sec. 5-3-5).

6-5 Show that Eq. (6-23) is valid.

BIBLIOGRAPHY

Chandrasekhar, S.: "Plasma Physics," The University of Chicago Press, Chicago, 1960.
sec. 1.6—The Relation between the Particle Orbit Theory and the Boltzmann Equation.
chap. 2—The First Order Orbit Theory.

Longmire, C. L.: "Elementary Plasma Physics," Interscience Publishers, Inc., New York, 1963.
chap. 2—First-order Orbit Theory.
chap. 3—Application of Orbit Theory to Static Problems.
chap. 4—Application of Orbit Theory to Dynamic Problems.

Northrop, T. G.: "The Adiabatic Motion of Charged Particles," Interscience Publishers, Inc., New York, 1963.
Contains a rigorous development of orbit theory.

chapter

7

The Interaction of Electromagnetic Waves with Plasmas

7-1 Introduction

When an electromagnetic wave encounters a plasma, the properties of the electromagnetic wave may be considerably altered. If this is the case, the properties of the plasma will also be altered. If the form of the interaction is understood and controlled, the results of the interaction can be useful to man. The electromagnetic field may contain the plasma in a finite region. This is, as we know, the aim of the controlled-fusion program. The development of microwave plasma tubes depends upon a detailed knowledge of the plasma–electromagnetic wave interaction. The results of a plasma–electromagnetic wave interaction can supply valuable information about the plasma and about the internal behavior of the ions and the neutral particles which are present in the plasma. On the other side of the coin, the plasma-induced distortion of a communication-carrying electromagnetic wave may lead to disaster. It is evident, in any event, that the plasma–electromagnetic wave interaction is of the utmost importance. We shall consider the fundamental aspects of that interaction.

7-2 Equations and Analysis

Our method of approach will be to solve Maxwell's equations in conjunction with the momentum transfer equations. We shall ignore the pressure term in the momentum transfer equations and shall consider a *constant* effective collision frequency. Since there is a static magnetic field present in most thermonuclear control machines and in atmospheric propagation (the earth's field), we shall consider a static uniform magnetic field in our formulation.

An electromagnetic wave in free space is composed of traveling electric and magnetic fields. The electric and magnetic fields are in time phase and perpendicular to each other in space. The ratio of the electric field intensity to the magnetic field intensity is given by

$$\frac{E}{H} = \left(\frac{\mu_0}{\epsilon_0}\right)^{1/2} \tag{7-1}$$

The magnitude of the force on a charged particle due to the electric field of the wave is $F_E = qE$, where q is either $-|e|$ for the electrons or $+Z_i|e|$ for the ions. The maximum magnitude of the force on a charged particle in the magnetic field is $F_B = quB$. The ratio of the electric to the maximum magnetic force is therefore

$$\frac{F_E}{F_B} = \frac{E}{u\mu_0 H} \tag{7-2}$$

or, using Eq. (7-1),

$$\frac{F_E}{F_B} = \frac{1}{u\sqrt{\mu_0\epsilon_0}} \tag{7-3}$$

Since the speed of light in vacuum c is equal to $1/\sqrt{\mu_0\epsilon_0}$, Eq. (7-3) can be written

$$\frac{F_E}{F_B} = \frac{c}{u} \tag{7-4}$$

For particle velocities considerably less than the speed of light, it is a good approximation to ignore the force on the particle due to the time-varying magnetic field of the electromagnetic wave in comparison with the force due to the electric field. This we shall do.

We wish, then, to solve the momentum transfer equations for the case of a propagating electric field and a static uniform magnetic field in the presence of a constant effective collision frequency. In Chap. 5, we solved the momentum transfer equations for the case of a time-varying spatially uniform electric field and a static uniform magnetic field in the presence of a constant effective collision frequency. We found that

$$\mathbf{j} = \langle \boldsymbol{\sigma} \rangle \cdot \mathbf{E} \tag{5-75}$$

where the conductivity tensor $\langle\boldsymbol{\sigma}\rangle$ is given by

$$\langle\boldsymbol{\sigma}\rangle = \frac{Z_i^2|e|^2 n_i}{m_i}\begin{bmatrix} \dfrac{\nu_{ci} - i\omega}{(\nu_{ci} - i\omega)^2 + \omega_{bi}^2} & \dfrac{\omega_{bi}}{(\nu_{ci} - i\omega)^2 + \omega_{bi}^2} & 0 \\ \dfrac{-\omega_{bi}}{(\nu_{ci} - i\omega)^2 + \omega_{bi}^2} & \dfrac{\nu_{ci} - i\omega}{(\nu_{ci} - i\omega)^2 + \omega_{bi}^2} & 0 \\ 0 & 0 & \dfrac{1}{\nu_{ci} - i\omega} \end{bmatrix}$$

$$+ \frac{|e|^2 n_e}{m_e}\begin{bmatrix} \dfrac{\nu_{ce} - i\omega}{(\nu_{ce} - i\omega)^2 + \omega_{be}^2} & \dfrac{\omega_{be}}{(\nu_{ce} - i\omega)^2 + \omega_{be}^2} & 0 \\ \dfrac{-\omega_{be}}{(\nu_{ce} - i\omega)^2 + \omega_{be}^2} & \dfrac{\nu_{ce} - i\omega}{(\nu_{ce} - i\omega)^2 + \omega_{be}^2} & 0 \\ 0 & 0 & \dfrac{1}{\nu_{ce} - i\omega} \end{bmatrix} \tag{5-76}$$

The electron and ion cyclotron frequencies are, respectively,

$$\omega_{be} = -\frac{|e|B_{0z}}{m_e} \qquad \text{and} \qquad \omega_{bi} = \frac{Z_i|e|B_{0z}}{m_i}$$

where B_{0z} is the static uniform magnetic field in the z direction. We shall assume that Eqs. (5-75) and (5-76) are valid for the case of a propagating electric field. The validity of this assumption will be discussed in Sec. 7-5.

Since we are considering a time-varying electromagnetic wave, the variable fields in Maxwell's equations can be assumed to vary as

$$\exp\,[i(\mathbf{k}\cdot\mathbf{r} - \omega t)]$$

where $\mathbf{k}$ is the propagation constant whose direction indicates the direction of propagation and whose magnitude equals 2π divided by the wavelength of the wave. From a knowledge of the frequency and propagation constant of the wave, the phase and group velocities in the direction of propagation of the wave can be determined.

$$v_{\text{phase}} = \frac{\omega}{k} \tag{7-5}$$

$$v_{\text{group}} = \frac{\partial\omega}{\partial k} \tag{7-6}$$

The determination of $\mathbf{k}$ as a function of ω will be one of our primary objectives.

After assuming that the variable field quantities in Maxwell's equations vary as exp $[i(\mathbf{k} \cdot \mathbf{r} - \omega t)]$ and taking the appropriate time and space derivatives, we can express Eqs. (4-1*a*) to (4-4*a*) as

$$i\epsilon_0\mathbf{k} \cdot \mathbf{E} = \rho \tag{7-7}$$
$$\mathbf{k} \times \mathbf{E} = \mu_0\omega\mathbf{H} \tag{7-8}$$
$$i\mathbf{k} \times \mathbf{H} = \mathbf{j} - i\epsilon_0\omega\mathbf{E} \tag{7-9}$$
$$\mu_0\mathbf{k} \cdot \mathbf{H} = 0 \tag{7-10}$$

if we assume that $\mathbf{D} = \epsilon_0\mathbf{E}$ and $\mathbf{B} = \mu_0\mathbf{H}$. Since $\mathbf{j} = \langle \boldsymbol{\sigma} \rangle \cdot \mathbf{E}$, the right-hand side of Eq. (7-9) may be written

$$\langle \boldsymbol{\sigma} \rangle \cdot \mathbf{E} - i\epsilon_0\omega\mathbf{E} \equiv -i\omega\langle \boldsymbol{\varepsilon} \rangle \cdot \mathbf{E} \tag{7-11}$$

Equation (7-11) defines the permittivity tensor $\langle \boldsymbol{\varepsilon} \rangle$.

$$\langle \boldsymbol{\varepsilon} \rangle = \epsilon_0 \left(\langle \mathbf{1} \rangle - \frac{\langle \boldsymbol{\sigma} \rangle}{i\epsilon_0\omega} \right) \tag{7-12}$$

where $\langle \mathbf{1} \rangle$ is a diagonal tensor whose elements are equal to unity. Equation (7-9) can then be rewritten

$$\mathbf{k} \times \mathbf{H} = -\omega\langle \boldsymbol{\varepsilon} \rangle \cdot \mathbf{E} \tag{7-13}$$

There must be a relationship between the current density, which we know as a function of the fields, and the charge density—a continuity equation for charge. To determine this relationship, let us dot $\mathbf{k}$ into both sides of Eq. (7-9). Since $\mathbf{k} \cdot \mathbf{k} \times \mathbf{H}$ is zero, Eq. (7-9) becomes

$$0 = \mathbf{k} \cdot \mathbf{j} - i\epsilon_0\omega\mathbf{k} \cdot \mathbf{E} \tag{7-14}$$

After substituting $\mathbf{k} \cdot \mathbf{E}$ from Eq. (7-7) into Eq. (7-14), we obtain

$$\rho = \frac{\mathbf{k} \cdot \mathbf{j}}{\omega} \tag{7-15}$$

the relation between ρ and $\mathbf{j}$. Equation (7-15) can also be written

$$\rho = \frac{\mathbf{k} \cdot \langle \boldsymbol{\sigma} \rangle \cdot \mathbf{E}}{\omega} \tag{7-16}$$

If we substitute ρ from Eq. (7-16) into Eq. (7-7), we find that

$$\mathbf{k} \cdot \langle \boldsymbol{\sigma} \rangle \cdot \mathbf{E} - i\epsilon_0 \omega \mathbf{k} \cdot \mathbf{E} = 0 \tag{7-17}$$

From a comparison of Eq. (7-17) with Eq. (7-11), it is apparent that

$$\mathbf{k} \cdot \langle \boldsymbol{\varepsilon} \rangle \cdot \mathbf{E} = 0 \tag{7-18}$$

Equation (7-18) contains information from Eqs. (7-7) and (7-9). We could have obtained Eq. (7-18) much more easily, but in not nearly so instructive a fashion, by dotting **k** into both sides of Eq. (7-13). Notice also that Eq. (7-10) can be obtained by dotting **k** into both sides of Eq. (7-8).

We can rewrite Maxwell's equations (7-7) to (7-10) in the form

$$\mathbf{k} \cdot \langle \boldsymbol{\varepsilon} \rangle \cdot \mathbf{E} = 0 \tag{7-19}$$
$$\mathbf{k} \times \mathbf{E} = \mu_0 \omega \mathbf{H} \tag{7-20}$$
$$\mathbf{k} \times \mathbf{H} = -\omega \langle \boldsymbol{\varepsilon} \rangle \cdot \mathbf{E} \tag{7-21}$$
$$\mathbf{k} \cdot \mathbf{H} = 0 \tag{7-22}$$

where the permittivity tensor is found from Eqs. (7-12) and (5-76). The plasma current density and the plasma charge density can be determined from Eqs. (5-75) and (7-16), respectively. Equations (7-19) to (7-22) specify the properties of the electromagnetic wave as it interacts with the plasma we are considering.

From the right-hand side of Eq. (7-21) we can determine the propagation constant as a function of the frequency of the electromagnetic wave if the medium in which the wave travels is isotropic. In an isotropic dielectric medium other than free space, the phase velocity of an electromagnetic wave is given by $v_p = 1/\sqrt{\mu_0 \epsilon}$. It follows, then, that $v_p/c = \sqrt{\epsilon_0/\epsilon}$ or $v_p/c = 1/\sqrt{K}$, where K is the dielectric constant of the medium. Since, in the direction of propagation, $k = \omega/v_p$, the magnitude of the propagation constant k can be written in terms of the dielectric constant

$$k = \frac{\omega}{c}\sqrt{K} \tag{7-23}$$

In the course of our investigation of plasma–electromagnetic wave interactions, we shall calculate the *effective* dielectric constant of the plasma K_p for several special cases. The application of Eq. (7-23) will then give us the dispersion relation, the relation between **k** and ω. Let us look now at four examples of the plasma–electromagnetic wave interaction.

7-3 Results and Conclusions

7-3-1 Plasma Propagation in the Absence of a Static Magnetic Field. $\nu_c = 0$. If $B_{0z} = 0$, the cyclotron frequencies ω_{bi} and ω_{be} are zero, and, in the absence of collisions, the permittivity tensor reduces to a diagonal tensor whose elements are equal. Thus the permittivity tensor reduces to a scalar *effective* permittivity.

$$\langle \varepsilon \rangle \rightarrow \epsilon_p = \epsilon_0 \left(1 - \frac{\omega_{pi}^2}{\omega^2} - \frac{\omega_{pe}^2}{\omega^2}\right) \tag{7-24}$$

where $\omega_{pi}^2 = \dfrac{Z_i^2 |e|^2 n_i}{m_i \epsilon_0}$ and $\omega_{pe}^2 = \dfrac{|e|^2 n_e}{m_e \epsilon_0}$

The quantities ω_{pi} and ω_{pe} are called the "ion plasma frequency" and the "electron plasma frequency," respectively. Since m_i is much greater than m_e (ions are heavier than electrons), the second term on the right of Eq. (7-24) can be ignored. Equation (7-24) yields the following scalar effective dielectric constant for the plasma:

$$K_p = \frac{\epsilon_p}{\epsilon_0} \cong 1 - \frac{\omega_{pe}^2}{\omega^2} \tag{7-25}$$

The propagation constant, from Eq. (7-23), becomes

$$k \cong \frac{\omega}{c}\left(1 - \frac{\omega_{pe}^2}{\omega^2}\right)^{1/2} \tag{7-26}$$

If the frequency of the electromagnetic wave is greater than the electron plasma frequency, k is a real number. The phase and group velocities of the wave can be determined from Eqs. (7-5) and (7-6). These are shown in Fig. 7-1. At frequencies for which $\omega \gg \omega_{pe}$, the propagation constant approaches its free-space value $k = \omega/c$, and the wave acts as if the plasma were not present. If the frequency of the electromagnetic wave is less than the electron plasma frequency, k is an imaginary number. Since the fields vary as exp $(i\mathbf{k} \cdot \mathbf{r})$, the phase-variation term will become a decaying exponential; that is, the wave will not propagate but will decay as it goes forward into the plasma. If an electromagnetic wave encounters a plasma under these circumstances, the wave will be reflected. Thus the ionosphere will reflect electromagnetic radiation whose frequency is less than the plasma frequency of the ionosphere.

This seemingly strange wave-plasma behavior can be explained as

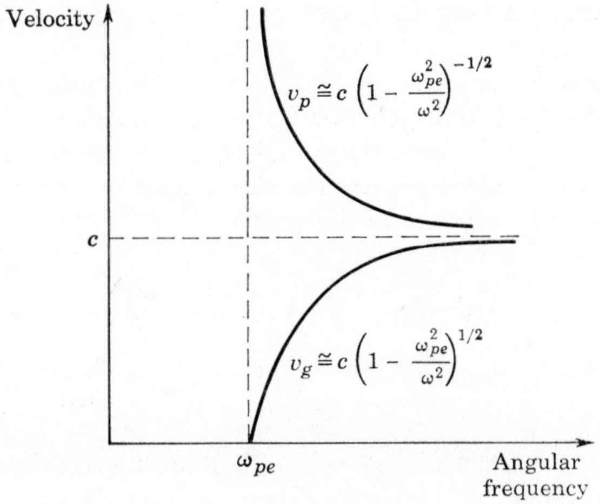

Fig. 7-1 Phase velocity and group velocity versus angular frequency for propagation in the absence of a static magnetic field and of collisions.

follows: From Eq. (7-9) we see that the real current density, the current density due to the motion of the charged plasma particles, is 180° out of phase with the displacement current. At very high frequencies the real current density is small compared with the displacement current density since the inertia of the plasma particles inhibits their motion in the rapidly changing electric field of the electromagnetic wave. The electromagnetic wave behaves as if it were in free space. At very low frequencies, the real current density is greater than the displacement current density. The sign of the total current density is reversed from the sign it would have in free space. The mathematical result of this sign reversal is that k becomes imaginary. The critical frequency for which the total current density becomes zero is the electron plasma frequency. As we shall see in Chap. 11, the electron plasma frequency is the natural frequency of oscillation of a system of plasma electrons. Thus it is not surprising that a discontinuity in the propagation characteristic occurs when the frequency of the electromagnetic wave equals the electron plasma frequency.

We can also explain the plasma-wave interaction from a slightly more physical point of view. At very low frequencies, the plasma particles have sufficient time to rearrange themselves so that they screen the interior of the plasma from the oscillating electric field of the electromagnetic wave. At high frequencies, the plasma particles do not have time to move between changes in the direction of the alternating electric field, so they can afford no screening.

The phase velocity of the electromagnetic wave is always greater than the speed of light in vacuum. The group velocity of the electromagnetic wave is always less than the velocity of light in vacuum. The group velocity can be considered as the speed of transmission of an electromagnetic signal through the dispersive plasma if the frequency of the wave is not close to the electron plasma frequency.

7-3-2 Plasma Propagation Perpendicular to a Static Magnetic Field. $E_y = E_x = 0$, $\nu_c = 0$. We now consider the case for which the electric field vector of the electromagnetic wave is parallel to the static uniform magnetic field. If we dot the permittivity tensor into an electric field with only a z component, the result, in the absence of collisions, is

$$\langle \boldsymbol{\varepsilon} \rangle \cdot \begin{pmatrix} 0 \\ 0 \\ E_z \end{pmatrix} = \epsilon_0 \left(1 - \frac{\omega_{pi}^2}{\omega^2} - \frac{\omega_{pe}^2}{\omega^2} \right) E_z \mathbf{a}_z \tag{7-27}$$

The effective dielectric constant is identical with that for the case of no static magnetic field [Eq. (7-24)]. In other words, if the electric field vector is in the same direction as the static magnetic field, the electromagnetic wave behaves as if the magnetic field were not present. The reason for this is as follows: For the static magnetic field to exert a force on a charged particle fluid, the velocity of that fluid must have some component in a direction perpendicular to the magnetic field. If the electric and magnetic fields are collinear, the average velocity of the charged particles will always remain parallel to the magnetic field. Thus there is never a magnetic force on the particle fluid, and we should not expect the magnetic field to appear in the expression for the effective dielectric constant of the plasma.

7-3-3 Plasma Propagation in the Direction of a Static Magnetic Field. $E_z = 0$, $\nu_c = 0$. We shall now consider the case in which the electric field of the time-varying wave is perpendicular to the static magnetic field and propagation is in the direction of the magnetic field. We shall again ignore collisions. An effective dielectric constant for the plasma can no longer be found. To gain detailed information about this type of propagation, we must solve Eqs. (7-20) and (7-21) simultaneously. Equations (7-20) and (7-21) can be written in component form as follows:

$$k_z E_y + \omega\mu_0 H_x = 0 \tag{7-28}$$
$$k_z E_x - \omega\mu_0 H_y = 0 \tag{7-29}$$
$$-A\omega E_x - GE_y + k_z H_y = 0 \tag{7-30}$$
$$-GE_x + A\omega E_y + k_z H_x = 0 \tag{7-31}$$

where
$$A = \epsilon_0 \left(1 + \frac{\omega_{pi}^2}{\omega_{bi}^2 - \omega^2} + \frac{\omega_{pe}^2}{\omega_{be}^2 - \omega^2}\right)$$

and
$$G = i\epsilon_0 \left(\frac{\omega_{bi}\omega_{pi}^2}{\omega_{bi}^2 - \omega^2} + \frac{\omega_{be}\omega_{pe}^2}{\omega_{be}^2 - \omega^2}\right)$$

In order that this homogeneous set of equations for the field quantities have other than a trivial solution, it is necessary that the following determinant vanish:

$$\begin{Vmatrix} 0 & k_z & \omega\mu_0 & 0 \\ k_z & 0 & 0 & -\mu_0\omega \\ -A\omega & -G & 0 & k_z \\ -G & A\omega & k_z & 0 \end{Vmatrix} = 0 \tag{7-32}$$

Equation (7-32) can be expanded to yield

$$k_z^4 - 2k_z^2\omega^2\mu_0 A + \omega^2\mu_0^2(\omega^2 A^2 + G^2) = 0 \tag{7-33}$$

which can in turn be solved for the propagation constant

$$k_z^2 = \omega^2\mu_0\epsilon_0 \left[1 - \frac{\omega_{pi}^2}{\omega(\omega \pm \omega_{bi})} - \frac{\omega_{pe}^2}{\omega(\omega \pm \omega_{be})}\right] \tag{7-34}$$

By virtue of the plus or minus signs in Eq. (7-34), the wave can assume two different velocities for a given frequency. Thus for propagation in the direction of the static magnetic field, the plasma acts like an anisotropic double refracting crystal in that there are two modes of propagation with two distinct velocities. Examination of Eq. (7-34) reveals that the propagation constant will exhibit several discontinuities with frequency. There will be "pass" and "stop" bands for the electromagnetic wave.

The unusual characteristics of the wave-plasma interaction in the presence of a static uniform magnetic field can be explained from a physical point of view. As we have previously noted, the charged particles of the plasma will attempt to move in the electric field of the electromagnetic wave in such a way that they screen the plasma from the electric field. These effects become evident in their pure form as ω_{bi} and ω_{be} become small compared with ω. Under these conditions, Eq. (7-34) reduces to Eq. (7-26). In addition to the screening effects ($k_z \to 0$), the plasma-wave interaction is characterized by "resonance" effects ($k_z \to \infty$). These effects are due to the interaction of the charged plasma particles with both the electromagnetic wave and the static magnetic field. In the absence of an electromagnetic wave, the ions and the electrons will orbit the static magnetic field in the directions shown in Fig. 7-2. The average ion velocity and the average electron velocity will be zero since the phases of the orbital motion of the ions and of the orbital motion of the electrons will, in general, be random. The time-varying electric field of the electromagnetic wave will alter the circular trajectories of the charged particles in a manner which is dependent upon the relationship between the frequency of the electromagnetic wave and the cyclotron frequencies of the charged particles. In any event, the average particle velocities induced by the time-varying electric field will be periodic with the frequency of the electromagnetic wave. We can consider the electromagnetic plane wave which we are discussing as being composed of two waves, one right circularly polarized (polarized clockwise, looking in the direction of propagation) and the other left circularly polarized. The two polarized waves are shown in Fig. 7-2. Notice that the static magnetic field and the direction of propagation are into the page. The properties of the two circularly polarized waves are determined by their interactions with the plasma particles. Since each of the two circularly polarized waves encounters a different type of particle behavior, it is to be expected that the characteristics

Fig. 7-2 Circularly polarized waves and orbiting charged particles.

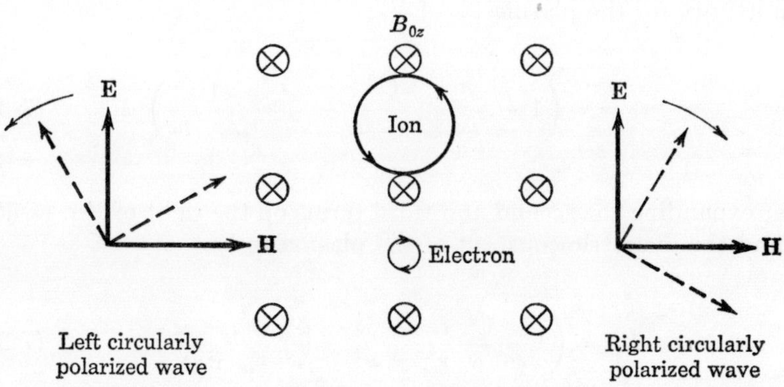

of the two circularly polarized waves will differ. We can, therefore, interpret physically the two modes of propagation which we have derived mathematically: Each mode of propagation corresponds to one of the circularly polarized waves. A resonance in the propagation properties of each circularly polarized wave occurs when the frequency of the wave equals the appropriate cyclotron frequency. Under these circumstances, the wave is "pumping" an oscillating system (one type of orbiting particles) at the resonant frequency of the system. The $\omega = \omega_{bi}$ resonance is caused by the left circularly polarized wave since that wave rotates in the direction in which the ions would like to orbit the static magnetic field, and hence the left circularly polarized wave is represented by the minus signs in Eq. (7-34). The right circularly polarized wave is represented by the plus signs in Eq. (7-34). (Note that the electron cyclotron frequency is defined as a negative quantity.)

It is instructive to compare the propagation "constant" for the type of propagation we have just discussed with that of Sec. 7-3-2. It is evident that the properties of the wave are a function of the direction of propagation. We have investigated the properties of the electromagnetic wave in the presence of a collisionless plasma and a static magnetic field for two special cases: (1) propagation in the direction of the static magnetic field with the electric field perpendicular to the static magnetic field and (2) propagation perpendicular to the static magnetic field with the electric field parallel to the static magnetic field. In between these two extremes there is a wide range of variation in the properties of the electromagnetic wave.

7-3-4 Plasma Propagation with Collisions. $B_{0z} = 0$. We shall now ignore the static magnetic field but allow collisions. The solution to be found is applicable to electromagnetic radiation interacting with metallic solids as well as with plasmas. The effective dielectric permittivity for the plasma is

$$\epsilon_p = \epsilon_0 \left(1 - \frac{\omega_{pi}^2}{i\omega} \frac{1}{\nu_{ci} - i\omega} - \frac{\omega_{pe}^2}{i\omega} \frac{1}{\nu_{ce} - i\omega} \right) \tag{7-35}$$

After expanding the second and third terms on the right of Eq. (7-35), we find the dielectric constant of the plasma to be

$$K_p = \left(1 + \frac{\omega_{pi}^2}{\omega} \frac{i\nu_{ci} - \omega}{\nu_{ci}^2 + \omega^2} + \frac{\omega_{pe}^2}{\omega} \frac{i\nu_{ce} - \omega}{\nu_{ce}^2 + \omega^2} \right) \tag{7-36}$$

and therefore the propagation constant is given by

$$k = \frac{\omega}{c}\left[1 - \frac{\omega_{pi}^2}{\nu_{ci}^2 + \omega^2} - \frac{\omega_{pe}^2}{\nu_{ce}^2 + \omega^2} + \frac{i\nu_{ci}\omega_{pi}^2}{\omega(\nu_{ci}^2 + \omega^2)} + \frac{i\nu_{ce}\omega_{pe}^2}{\omega(\nu_{ce}^2 + \omega^2)}\right]^{1/2} \tag{7-37}$$

The fourth and fifth terms on the right of Eq. (7-37) lead to an expression for the "skin depth" of the wave, the length into the plasma at which the wave is damped to about 1/2.7 of its original value by collision processes. The energy of the electromagnetic wave is transferred to the electrons and ions which dissipate this acquired energy in collisions.

7-4 The General Approach

In order to determine the behavior of an electromagnetic wave in the presence of a plasma, one must solve Maxwell's equations for the assumed type of interaction. It is convenient to combine Maxwell's equations into a wave equation which can then be solved to yield the same information. We shall now derive the wave equation for plasma propagation.

If we cross $\mathbf{k}$ into both sides of Eq. (7-20), we find that Eq. (7-20) becomes

$$\mathbf{k} \times \mathbf{k} \times \mathbf{E} = \omega\mu_0\mathbf{k} \times \mathbf{H} \tag{7-38}$$

Equation (7-38) can be combined with Eq. (7-21) to yield

$$\mathbf{k} \times \mathbf{k} \times \mathbf{E} + \omega^2\mu_0\langle\boldsymbol{\varepsilon}\rangle \cdot \mathbf{E} = 0 \tag{7-39}$$

Equation (7-39) is the wave equation for the anisotropic medium.

By expanding $\mathbf{k} \times \mathbf{k} \times \mathbf{E}$ and then collecting the terms, it is possible to show that

$$\mathbf{k} \times \mathbf{k} \times \mathbf{E} = \langle\boldsymbol{\Lambda}\rangle \cdot \mathbf{E} \tag{7-40}$$

where the tensor $\langle\boldsymbol{\Lambda}\rangle$ is defined as

$$\langle\boldsymbol{\Lambda}\rangle \equiv \begin{bmatrix} -k_y^2 - k_z^2 & k_xk_y & k_xk_z \\ k_yk_x & -k_z^2 - k_x^2 & k_yk_z \\ k_zk_x & k_zk_y & -k_x^2 - k_y^2 \end{bmatrix} \tag{7-41}$$

The wave equation (7-39) can therefore be written

$$(\langle \mathbf{\Lambda} \rangle + \omega^2 \mu_0 \langle \boldsymbol{\varepsilon} \rangle) \cdot \mathbf{E} = 0 \tag{7-42}$$

The solution to this equation is found by setting the determinant of the second-order tensor which multiplies **E** equal to zero.

$$\| (\langle \mathbf{\Lambda} \rangle + \omega^2 \mu_0 \langle \boldsymbol{\varepsilon} \rangle) \| = 0 \tag{7-43}$$

The vector propagation constant can be found as a function of frequency from Eq. (7-43). This result can be inserted into Eq. (7-42), which can then be solved for the electric field. Once the electric field and the propagation constant are known, the magnetic field can be calculated from Eq. (7-20). This method of approach can be used to predict the occurrence and behavior of a wide variety of plasma waves.

7-5 Assumptions and Consequences

We have made a number of assumptions in the foregoing sections. The results of our discussion should not be applied to situations in which these assumptions are not warranted. This is obvious but well worth emphasizing.

We have assumed that the static magnetic field is uniform. In practice it is impossible to create a uniform magnetic field. The calculations will therefore not reveal plasma phenomena that occur because the magnetic field is nonuniform.

We have assumed that the speed of the charged particles is at least an order of magnitude less than the speed of light so that the force on the particles due to the time-varying magnetic field is negligible. This is a good approximation for electron energies less than 10^4 ev and for ion energies less than 10^7 ev.

We have assumed that the effective collision frequencies are not a function of particle average velocities or particle densities, that is, that they are constant. This is often a poor assumption. If, for instance, we allow the effective collision frequencies to be functions of particle average velocities, the collision terms in the momentum transfer equations become nonlinear. This is the nonlinearity which causes the Luxembourg effect. The electron effective collision frequency is modulated by one wave, and this modulation of effective collision frequency changes the characteristics of the second wave. In general, the collision-frequency terms are nonlinear. There are, however, certain situations for which the effective collision frequencies are essentially constant (see Prob. 5-2 and Sec. 10-2-5).

We have assumed that the pressure gradients in the momentum transfer equations can be neglected. Pressure is a measure of the random thermal energy density of a group of particles. If the directional energy induced by external fields is much larger than the random thermal energy of the particles, the thermal energy and thus the pressure can be ignored. This is sometimes called the "$T = 0°\text{K}$ approximation" since thermal pressure is zero at zero temperature. Even if the pressure is appreciable, the pressure gradient may still be negligible. This may be the case if the applied fields have little or no effect on the particle densities.

We have made another assumption which should be discussed. We have assumed that the real dielectric permittivity of the plasma is ϵ_0 and that the real permeability of the plasma is μ_0. If there are ions which are not fully ionized in the plasma or neutral particles in the plasma, the electric dipole moments of these ions and neutral particles increase the value of the real dielectric constant from the value of unity which it has in vacuum (although this increase is usually negligible). Therefore, if the plasma is not fully ionized, the vacuum permittivity which we used in Maxwell's equations must be replaced by the real dielectric permittivity of the plasma. An analogous discussion regarding the use of μ, the real permeability, instead of μ_0 may be advanced.

Lastly, we have assumed that the conductivity tensor derived in Chap. 5 is valid for the case that the particle average velocities and the electric fields vary as $\exp\,[i(\mathbf{k}\cdot\mathbf{r} - \omega t)]$. Let us assume that the particle average velocities do vary as $\exp\,[i\,(\mathbf{k}\cdot\mathbf{r} - \omega t)]$. The time derivatives in the momentum transfer equations can be expanded as

$$\frac{d\mathbf{v}}{dt} = \frac{\partial \mathbf{v}}{\partial t} + \frac{\partial \mathbf{v}}{\partial x}\frac{dx}{dt} + \frac{\partial \mathbf{v}}{\partial y}\frac{dy}{dt} + \frac{\partial \mathbf{v}}{\partial z}\frac{dz}{dt} \tag{7-44}$$

If we make use of the assumed time and spatial variation, Eq. (7-44) can be written

$$\frac{d\mathbf{v}}{dt} = -i\omega\mathbf{v} + i(\mathbf{k}\cdot\mathbf{v})\mathbf{v} \tag{7-45}$$

If the average velocities of the particles are always perpendicular to the direction of propagation, $\mathbf{k}\cdot\mathbf{v}$ is always equal to zero. The resulting expressions for the momentum transfer equations and for the conductivity tensor are the same as the expressions obtained if the particle average velocities and the electric fields are assumed to vary as $\exp(-i\omega t)$. For the examples we have investigated in Sec. 7-3, the particle average velocities are perpendicular to the direction of propagation, and thus Eqs. (5-75) and (5-76) are valid.

For the general case in which particle average motion is not perpendicular to the direction of propagation, the velocity-squared term on the right-hand side of Eq. (7-45) causes the momentum transfer equations to be nonlinear. If it is assumed that the average velocity is small, the term in velocity squared can be ignored. Thus the conductivity tensor which we derived in Chap. 5 can be used. In the general case, therefore, if we wish to encounter linear equations, it is necessary to make the "small-velocity" approximation. In practice it is almost always necessary to make this approximation in order to obtain a solution to the momentum transfer equations.

PROBLEMS

7-1 Show that if $\mathbf{A} \sim \exp[i(\mathbf{k}\cdot\mathbf{r} - \omega t)]$, then

$$\nabla \cdot \mathbf{A} = i\mathbf{k}\cdot\mathbf{A}$$
$$\nabla \times \mathbf{A} = i\mathbf{k}\times\mathbf{A}$$
$$\frac{d\mathbf{A}}{dt} = -i\omega\mathbf{A} + i(\mathbf{k}\cdot\mathbf{v})\mathbf{A}$$

where
$$\mathbf{v} = \frac{d\mathbf{r}}{dt}$$

7-2 Find analytical expressions for the phase velocity and the group velocity of an electromagnetic wave propagating in a collisionless plasma in the absence of a static magnetic field.

7-3 Show that Eq. (7-34) follows from Eq. (7-32).

7-4 Plot k_z versus ω from Eq. (7-34) for both modes of propagation if (*a*) $\omega_{be} = 10\omega_{pe}$ and (*b*) $\omega_{pe} = 10\omega_{be}$. Using these graphs of k_z versus ω, plot the phase and group velocities of the electromagnetic waves as a function of ω.
RECALL:

$$\omega_{be} = -\frac{|e|B_{0z}}{m_e} \qquad \omega_{bi} = +\frac{Z_i|e|B_{0z}}{m_i}$$

7-5 Find and plot the skin depth of a plasma (electron effective collision frequency ν_{ce} and ion effective collision frequency ν_{ci}) as a function of the frequency of the impinging electromagnetic wave.

7-6 Show that Eq. (7-40) is valid.

7-7 Find the dispersion relation for propagation of an electromagnetic wave through a collisionless plasma immersed in a uniform static magnetic field if the direction of propagation is perpendicular to the magnetic field. Since there will be $\mathbf{E} \times \mathbf{B}$ drifts in the direction of propagation, it will be necessary to make the small-velocity approximation. Assume propagation in the y direction with the electric field in the x direction.

7-8 Qualitatively describe the trajectories of charged plasma particles in the presence of a uniform static magnetic field and an electromagnetic wave propagating in the direction of the magnetic field. Consider both the case in which the charged particles have no thermal energy and the case in which the charged particles have thermal energy. In each case describe the particle trajectories if $\omega > \omega_{bi}, \omega_{be}$; $\omega < \omega_{bi}, \omega_{be}$; $\omega = \omega_{bi}$; $\omega = \omega_{be}$.

BIBLIOGRAPHY

Allis, W. P., S. J. Buchsbaum, and A. Bers: "Waves in Anisotropic Plasmas," The M.I.T. Press, Cambridge, Massachusetts, 1963.
An advanced treatment of free and guided wave propagation in plasmas.

Astrom, E.: On Waves in an Ionized Gas, *Arkiv. Fysik,* **2**:443 (1950).
On the solution of Eq. (7-43).

Denisse, J. F., and J. L. Delcroix: "Plasma Waves," Interscience Publishers, Inc., New York, 1963.
An advanced treatment of all phases of wave-plasma interactions.

Francis, G.: "Ionization Phenomena in Gases," Butterworth Scientific Publications, London, 1960.
p. 271—On the solution of Eq. (7-43).

Hayt, W. H.: "Engineering Electromagnetics," McGraw-Hill Book Company, Inc., New York, 1958.
sec. 11.1—Wave Motion. On the solution of the wave equation in free space.

Mitra, S. K.: "The Upper Atmosphere," Royal Asiatic Society of Bengal, 1947.
p. 143—A discussion of the types of ionospheric propagation and their properties.

Ramo, S., and J. R. Whinnery: "Fields and Waves in Modern Radio," John Wiley & Sons, Inc., New York, 1953.
chap. 7—Propagation and Reflection of Electromagnetic Waves.

Rose, D. J., and M. Clark, Jr.: "Plasmas and Controlled Fusion," John Wiley & Sons, Inc., New York, 1961.
chap. 9—Plasma Waves of Small Amplitude.

Stix, T. H.: "The Theory of Plasma Waves," McGraw-Hill Book Company, Inc., New York, 1962.
An advanced treatment of plasma waves.

Stratton, J. A.: "Electromagnetic Theory," McGraw-Hill Book Company, Inc., New York, 1941.
p. 327—On propagation in an ionized atmosphere with a static magnetic field present.
p. 330—A discussion of group velocity, phase velocity, and signal velocity.

chapter

8

Magnetic and Kinetic Pressures

8-1 Introduction

The concept of a kinetic pressure, a pressure due to a group of particles in thermal motion, has been discussed in Chap. 3. In Chap. 5, we investigated the forces which charged particles experience when exposed to magnetic fields. Since force and pressure are related quantities, it would seem reasonable that a *magnetic pressure* could be defined. It is not at all obvious what form the magnetic pressure should take except that it must have the same dimensions as kinetic pressure. It is further not obvious that the concept of a magnetic pressure will be useful.

In this chapter we shall combine the momentum transfer equations for ions and for electrons in such a way that from the resulting expression we shall be able to define a magnetic pressure which is both mathematically and physically acceptable. We shall find that the defined magnetic pressure is a very useful qualitative concept in the discussion of the interaction of a plasma with a static magnetic field.

8-2 Derivation

We shall assume that we have been presented a *fully ionized* plasma ($n_n = 0$) in which there is a static magnetic field and a kinetic pressure gradient but in which the particle densities and velocities are not a function of time. In other words, the plasma is an equilibrium conglomeration of equal positive and negative charge densities ($Z_i n_i = n_e$). If we assume charge neutrality and total ionization, the addition of

$n_e m_e(d\mathbf{v}_e/dt)$ and $n_i m_i(d\mathbf{v}_i/dt)$ from Eqs. (4-10) and (4-11), respectively, yields

$$n_i m_i \frac{d\mathbf{v}_i}{dt} + n_e m_e \frac{d\mathbf{v}_e}{dt} = (Z_i|e|n_i\mathbf{v}_i - |e|n_e\mathbf{v}_e) \times \mathbf{B} - \nabla(P_i + P_e) \qquad (8\text{-}1)$$

Since there are no neutral particles present, the frictional force on the electron fluid must be equal and opposite in direction to the frictional force on the ion fluid. The frictional terms therefore add to zero. Because of the assumed charge neutrality, the terms in the electric field add to zero. The term crossed into **B** is the current density **j**, as given by Eq. (4-27). The total kinetic pressure P is the sum of the electron and ion pressures, $P = P_e + P_i$. Equation (8-1) can therefore be written

$$n_i m_i \frac{d\mathbf{v}_i}{dt} + n_e m_e \frac{d\mathbf{v}_e}{dt} = \mathbf{j} \times \mathbf{B} - \nabla P \qquad (8\text{-}2)$$

The average velocity of *mass* flow within the plasma is given by

$$\mathbf{U} = \frac{n_i m_i \mathbf{v}_i + n_e m_e \mathbf{v}_e}{n_i m_i + n_e m_e} \qquad (8\text{-}3)$$

If we assume charge neutrality and assume that $m_i \gg m_e$, Eqs. (8-2) and (8-3) can be combined to yield

$$n_i m_i \frac{d\mathbf{U}}{dt} = \mathbf{j} \times \mathbf{B} - \nabla P \qquad (8\text{-}4)$$

We shall assume that the mass flow is small. We can therefore make the small-velocity approximation: We can ignore second-order terms in **U** which appear when $d\mathbf{U}/dt$ is expanded.

$$\begin{aligned}\frac{d\mathbf{U}}{dt} &= \frac{\partial \mathbf{U}}{\partial t} + \frac{\partial \mathbf{U}}{\partial x}\frac{dx}{dt} + \frac{\partial \mathbf{U}}{\partial y}\frac{dy}{dt} + \frac{\partial \mathbf{U}}{\partial z}\frac{dz}{dt} \\ &= \frac{\partial \mathbf{U}}{\partial t} + \frac{\partial \mathbf{U}}{\partial x} U_x + \frac{\partial \mathbf{U}}{\partial y} U_y + \frac{\partial \mathbf{U}}{\partial z} U_z \\ &= \frac{\partial \mathbf{U}}{\partial t} + (\mathbf{U} \cdot \nabla)\mathbf{U} \cong \frac{\partial \mathbf{U}}{\partial t}\end{aligned}$$

We shall now assume that the particle densities and velocities are independent of time. Equation (8-4) reduces to a relation which expresses the balance between pressure and magnetic forces.

$$\mathbf{j} \times \mathbf{B} = \nabla P \qquad (8\text{-}5)$$

Notice that Eq. (8-5) can be derived directly from Eq. (8-1) by making the small-velocity approximation and assuming time-independent velocities. It is apparent from Eq. (8-5) that ∇P is perpendicular to $\mathbf{B}$ and that therefore the pressure must be constant along the magnetic field lines. The current density in the system is perpendicular to both the magnetic field and the pressure gradient.

If the time derivative in Maxwell's equation (4-2*a*) is set equal to zero, that equation becomes

$$\nabla \times \mathbf{H} = \mathbf{j} \tag{8-6}$$

Equations (8-6) and (8-5) can be combined to eliminate $\mathbf{j}$

$$(\nabla \times \mathbf{H}) \times \mathbf{B} = \nabla P$$

or, assuming $\mathbf{B} = \mu_0 \mathbf{H}$,

$$\frac{(\nabla \times \mathbf{B}) \times \mathbf{B}}{\mu_0} = \nabla P \tag{8-7}$$

The following vector identity will be employed to render $(\nabla \times \mathbf{B}) \times \mathbf{B}$ more tractable:

$$(\nabla \times \mathbf{B}) \times \mathbf{B} = (\mathbf{B} \cdot \nabla)\mathbf{B} - \tfrac{1}{2}\nabla|\mathbf{B}|^2$$

Equation (8-7) can therefore be written

$$\nabla\left(P + \frac{|\mathbf{B}|^2}{2\mu_0}\right) = \frac{(\mathbf{B} \cdot \nabla)\mathbf{B}}{\mu_0} \tag{8-8}$$

We shall consider the case for which $(\mathbf{B} \cdot \nabla)\mathbf{B}$ is zero: the straight magnetic field. For that case, Eq. (8-8) reduces to

$$\nabla\left(P + \frac{|\mathbf{B}|^2}{2\mu_0}\right) = 0 \tag{8-9}$$

The solution to Eq. (8-9) is

$$P + P_m = \text{constant} \tag{8-10}$$

where the magnetic pressure P_m is defined as

$$P_m \equiv \frac{|\mathbf{B}|^2}{2\mu_0} \tag{8-11}$$

The defined magnetic pressure has the dimensions of kinetic pressure. In addition, from an examination of the form of Eq. (8-9), we would expect the defined magnetic pressure to behave as if it were a real pressure.

We can calculate the constant on the right-hand side of Eq. (8-10) as follows: If the plasma is surrounded by a region in which there are no particles, a region where $P = 0$, the left-hand side of Eq. (8-10) evaluated in that region is equal to $|\mathbf{B}_0|^2/2\mu_0$, where $\mathbf{B}_0$ is the magnetic field just outside the plasma. The constant on the right-hand side of Eq. (8-10) must therefore equal $|\mathbf{B}_0|^2/2\mu_0$.

$$P + \frac{|\mathbf{B}|^2}{2\mu_0} = \frac{|\mathbf{B}_0|^2}{2\mu_0} \tag{8-12}$$

We can define a parameter β as the ratio of the kinetic pressure at a point within the plasma to the confining magnetic pressure at the edge of the plasma.

$$\beta \equiv \frac{P}{|\mathbf{B}_0|^2/2\mu_0} \tag{8-13}$$

The combination of Eqs. (8-12) and (8-13) yields

$$\beta = 1 - \frac{|\mathbf{B}|^2}{|\mathbf{B}_0|^2} \tag{8-14}$$

In the jargon of the controlled-fusion researchers, fusion machines can be either "high β" or "low β." If the kinetic pressure is low compared with the magnetic pressure at the boundary of the plasma, the machine is a "low β"; if the kinetic pressure is of the same order of magnitude as the magnetic pressure at the boundary of the plasma, the machine is a "high β" (β approaching unity).

If a magnetic field is straight, that magnetic field will confine a plasma whose pressure at all points is less than the magnetic pressure at the boundary of the plasma. Since no magnetic field is perfectly straight, the concept of a magnetic pressure is of value chiefly in a qualitative way. For a magnetic flux density of 0.5 weber/m^2, the corresponding magnetic pressure is about 1 atm. Thus a plasma at a kinetic pressure of about 1 atm can be contained by a magnetic field of 0.5 weber/m^2. The magnetic pressure, however, increases as the square of the magnetic flux density. It follows that a plasma whose kinetic pressure is about 100 atm can be confined by a magnetic flux density of about 5 webers/m^2. A controlled thermonuclear reaction can, from a theoretical point of view, operate at a kinetic pressure of

100 atm. A magnetic flux density of 5 webers/m^2 is a very large, but attainable, field.

8-3 Discussion

For the special case of straight magnetic field lines, the sum of the kinetic and magnetic pressures is a constant. Thus a plasma confined in a magnetic field is diamagnetic. The motion of the plasma particles sets up a magnetic field which opposes the original magnetic field. At points in the plasma where the kinetic pressure is greatest, the bucking field is also greatest, and the original magnetic field is decreased in magnitude. Physically this result is quite reasonable. Kinetic pressure is proportional to the particle densities and average squared velocities. Each individual particle, in orbiting around the original magnetic field, sets up a small bucking field. The total bucking field is equal to the sum of the small bucking fields from the individual particles and, therefore, is proportional to the kinetic pressure.

In our discussion of orbit theory we found that for straight magnetic field lines

$$W_\perp + \frac{|\mathbf{B}|^2}{2\mu_0} = \text{constant} \tag{6-18}$$

where $W_\perp = n_e\overline{\mathcal{E}_{e\perp}} + n_i\overline{\mathcal{E}_{i\perp}}$. It follows from comparing Eq. (6-18) with Eq. (8-10) that $W_\perp$ can be interpreted as the kinetic pressure of the plasma perpendicular to the magnetic field.

It is worth remembering that some of the results which we have obtained in this chapter are strictly valid only for magnetic field geometries in which $(\mathbf{B} \cdot \nabla)\mathbf{B} = 0$ and for situations in which the mass flow is small. Under other conditions these results may be valid, but such validity must be proved. Further, it has been implicitly assumed that the plasma ions have no intrinsic magnetic properties. Hence we have used the vacuum permeability μ_0 in our analysis. The parameter β which we have discussed is defined in some literature as the ratio of the kinetic pressure at a point within the plasma to the magnetic pressure at the *same point*. If this definition is used, β may assume values from zero to infinity.

PROBLEMS

8-1 Show that Eq. (8-4) follows from Eqs. (8-2) and (8-3) if it is assumed that $Z_i n_i = n_e$ and $m_i \gg m_e$.

8-2 Show that $(\mathbf{B} \cdot \nabla)\mathbf{B} \neq 0$ if $\mathbf{B} = |\mathbf{B}(r)|\mathbf{a}_\theta$, where $\mathbf{a}_\theta$ is the unit vector in the θ direction of a cylindrical coordinate system. Calculate $(\mathbf{B} \cdot \nabla)\mathbf{B}$ if $\mathbf{B} = (k/r)\mathbf{a}_\theta$, where k is a constant.

8-3 Determine the approximate magnetic flux density needed to confine a plasma whose particles have an average energy of 10^5 ev. Express the magnetic flux density as a function of the maximum particle densities present in the plasma. What particle density can be confined by a field of 1 weber/m²? Find the radius of an orbiting electron in the plasma under these conditions.

8-4 By subtracting $m_e^2 n_e(d\mathbf{v}_e/dt)$ from $(m_i^2 n_i/Z_i)(d\mathbf{v}_i/dt)$, show that in the equilibrium state in the absence of collisions the plasma mass flow perpendicular to a magnetic field is given by

$$\mathbf{U}_\perp = \frac{\mathbf{B}}{|\mathbf{B}|^2} \times \left(-\mathbf{E} + \frac{\nabla P_i}{|e|n_e}\right) \tag{8-15}$$

In order to obtain this result, it is necessary to use the small-velocity approximation and the condition of charge neutrality and to assume that $m_i \gg m_e$. For the case of straight magnetic field lines and a cylindrical plasma whose axis is along the magnetic field lines, draw a diagram showing the directions of $\mathbf{U}_\perp$, $\mathbf{E}$, ∇P, $\mathbf{j}$. Discuss.

8-5 If the electric field in Eqs. (4-10) and (4-11) is an induced electric field, that is, an electric field caused by a net charge density within the plasma, then $n_e \neq Z_i n_i$. Show formally that if $n_e \cong Z_i n_i$, then Eqs. (8-4) and (8-15) are valid first approximations. We shall show in Chap. 9 that within a plasma the electron and ion charge densities are approximately equal.

BIBLIOGRAPHY

Glasstone, S., and R. Lovberg: "Controlled Thermonuclear Reactions," D. Van Nostrand Company, Inc., Princeton, N.J., 1960.
p. 50—On the magnetic pressure and definition of β.

Rose, D. J., and M. Clark, Jr.: "Plasmas and Controlled Fusion," John Wiley & Sons, Inc., New York, 1961.
sec. 7.3—On the magnetic pressure and definition of β.

Spitzer, L., Jr.: "Physics of Fully Ionized Gases," 2d ed., Interscience Publishers, Inc., New York, 1962.
sec. 2.2—On the equations for the average velocity of mass flow.
sec. 4.2*a*—On the magnetic pressure.

chapter

9

Plasma Sheaths and the Debye Length

9-1 Introduction

If a plasma comes in contact with a material body, a sheath or contact potential region is formed between the plasma and the body. Across the sheath region there will be a voltage drop which is very similar in form to the contact potential appearing at a semiconductor-metal junction or at a *p-n* semiconductor junction.

The existence of the sheath region can be predicted by means of a physical argument. In order to be able to advance this physical argument, we must first prove that, in general, the electrons in a plasma have a much larger random speed than the positive ions. To show that this is the case, let us assume initially that the average energies of the electrons and of the ions are equal (we shall see soon that the electron average energy is often the greater).

$$\tfrac{1}{2}m_i\overline{u_i^2} = \tfrac{1}{2}m_e\overline{u_e^2} \tag{9-1}$$

Equation (9-1) can be solved for the ratio of the electron root mean square velocity to the ion root mean square velocity.

$$\left(\frac{\overline{u_e^2}}{\overline{u_i^2}}\right)^{1/2} = \left(\frac{m_i}{m_e}\right)^{1/2} \tag{9-2}$$

Since the lightest ion is about 1,836 times heavier than an electron, $(\overline{u_e^2}/\overline{u_i^2})^{1/2}$ is always greater than about 43.

We shall now consider the mechanism by which the electrons can obtain much higher average energies than the positive ions. If an electric field is present, the electrons gain energy from the field between collisions. The energy gained from the field is independent of the electrons' average energy. When an electron makes a collision, it loses some of its energy. The amount of energy an electron loses in a collision is often a function of its average energy. The fractional electron energy loss is usually very small. For example, if an electron collides elastically (kinetic energy being conserved) with a positive ion, the maximum fractional electron energy loss is $4m_e/m_i$. A ping-pong ball gives up very little energy in bouncing off a battleship. If the plasma is not in a transient state, the energy supplied per unit time to the electrons by the field must be transferred from the electrons to the ions and neutral particles by collisions. Only in this way can the electron average energy be time-invariant. Thus the condition for equilibrium can be expressed as

$$\overline{\mathfrak{V}} = \overline{F(\mathcal{E}_e)\mathcal{E}_e} \tag{9-3}$$

where $\overline{\mathfrak{V}}$ is the average energy gained by an electron from the electric field between collisions, $F(\mathcal{E}_e)$ is the fractional electron energy loss per collision, and $\overline{F(\mathcal{E}_e)\mathcal{E}_e}$ is the average electron energy loss per collision. $\overline{\mathfrak{V}}$ is a function of $|\mathbf{E}|/P$, where $\mathbf{E}$ is the electric field and P is the pressure, since the energy gained between collisions is directly proportional to the field strength and to the average distance between collisions (which is inversely proportional to P). Often for pressures below 10 mm Hg, Eq. (9-3) yields an average electron energy far in excess of the average ion energy. In a voltage regulator tube operating at 1 mm Hg, the electron temperature is about 10^4°K or about 1 ev, while the ion temperature is about 300°K or about .03 ev (temperature is defined by $\bar{\mathcal{E}} = \frac{3}{2}\kappa T$). We can conclude, then, that unless very special methods are employed to provide otherwise, the electron random velocity is always much greater than the ion random velocity. We are now prepared to resume our plausibility argument for the existence of a plasma sheath.

The random particle current density for electrons and the random particle current density for ions, respectively, in a plasma are

$$J_e = \frac{n_e\overline{u_e}}{4} \qquad J_i = \frac{n_i\overline{u_i}}{4}$$

Since $\overline{u_e} \gg \overline{u_i}$, the random particle current density for electrons is much greater than that for ions, $J_e \gg J_i$. If an isolated material body is brought into contact with a plasma, the electron current to the body is initially much greater than the ion current. The isolated body will therefore be charged negatively until its potential is sufficiently lowered that the net charge arriving at the body is zero; that is, since no current can flow through the isolated body, an equal number of positive and negative charges must arrive at the body per unit time for equilibrium to exist. Thus we see that there must be a potential difference, a contact potential, between the plasma and the material body. The isolated material body is always at a negative potential with respect to the plasma. The region across which the potential drop takes place, the sheath region, must repel some of the negative charge leaving the plasma. The positive ions which enter the sheath will be rapidly swept to the material body by the sheath electric field. It follows that the positive and negative charge densities within the sheath region will be unequal.

In the same way that a plasma isolates itself from a negatively charged material body, the plasma will isolate itself from an extraneous charge placed within the plasma; that is, surrounding the extraneous charge there will be a spherical region of unequal charge densities and of potential drop. The approximate radius of this spherical region is called the "Debye shielding distance" or the "Debye length." At more than one Debye length from any charge unbalance within the plasma, the plasma will be essentially unaffected by the charge unbalance. At less than one Debye length, sheath characteristics will prevail. Around an extraneous negative charge there will be a spherical region of net positive charge density. Around an extraneous positive charge there will be a spherical region of net negative charge density. It is in this way that the plasma strives to maintain internal charge neutrality. The struggle to maintain internal charge neutrality is a characteristic of the plasma state which differentiates that state from the other states of matter. One of the defining characteristics of a plasma is, therefore, that the Debye length be very small compared with the physical dimensions of the plasma. When mathematically formulated, this restriction will replace a part of the rather loose definition of the plasma state set forth in the opening paragraphs of Chap. 1.

We have used the word "extraneous" in the preceding paragraph to characterize a charge found in a region of the plasma which otherwise would be space-charge-neutral. In point of fact, every charge in the plasma can be considered an extraneous charge.

9-2 The Plasma Sheath

We shall consider the case of a plasma in contact with an isolated material body which lies in the $x = 0$ plane. We shall assume that equilibrium has been attained and that the potential of the body is V_0. Later we shall solve for V_0. We shall also assume that the potential of the plasma at $x = \infty$ is $V_\infty = 0$ and that the electron and ion densities are given by Maxwell-Boltzmann statistics:

$$n_e = n_{e\infty} \exp\left[\frac{|e|V(x)}{\kappa T_e}\right] \tag{9-4}$$

$$n_i = n_{i\infty} \exp\left[-\frac{Z_i|e|V(x)}{\kappa T_i}\right] \tag{9-5}$$

where $n_{e\infty}$ and $n_{i\infty}$ are the electron and ion densities at $x = \infty$, T_e is the electron temperature, and T_i is the ion temperature. Since $E_x = -\partial V(x)/\partial x$, Eq. (4-1*a*) can be written as Poisson's equation

$$\frac{\partial^2 V(x)}{\partial x^2} = -\frac{\rho}{\epsilon_0} \tag{9-6}$$

where the permittivity of the plasma is assumed to be that of free space. The charge density ρ is given by Eq. (4-26).

$$\rho = (Z_i|e|n_i - |e|n_e) \tag{4-26}$$

The combination of Eqs. (9-4) to (9-6) and (4-26) yields

$$\frac{\partial^2 V(x)}{\partial x^2} = \frac{-|e|}{\epsilon_0}\left\{Z_i n_{i\infty} \exp\left[-\frac{Z_i|e|V(x)}{\kappa T_i}\right] - n_{e\infty} \exp\left[\frac{|e|V(x)}{\kappa T_e}\right]\right\} \tag{9-7}$$

The solution of Eq. (9-7) will give the potential variation in the plasma and hence the effective sheath width. When we have determined the potential variation, the individual particle densities can be calculated from Eqs. (9-4) and (9-5). Unfortunately, Eq. (9-7) cannot be solved as it stands without the use of a computing machine. To solve Eq. (9-7) within the confines of this book, it is necessary to make the following approximation:

$$Z_i|e|V_0 < \kappa T_i \qquad |e|V_0 < \kappa T_e \tag{9-8}$$

We shall check the validity of this approximation later. The exponentials in Eq. (9-7) can then be expanded ($e^y \cong 1 + y$) to yield

$$\frac{\partial^2 V}{\partial x^2} = \left(\frac{Z_i^2|e|^2 n_{i\infty}}{\epsilon_0 \kappa T_i} + \frac{|e|^2 n_{e\infty}}{\epsilon_0 \kappa T_e}\right) V \tag{9-9}$$

If we assume that there is charge neutrality at infinity, the solution to Eq. (9-9), subject to the given boundary conditions, is

$$V(x) = V_0 e^{-x/X} \tag{9-10}$$

where the effective sheath width X is

$$X = \frac{1}{(Z_i|e|^2 n_{e\infty}/\epsilon_0 \kappa T_i + |e|^2 n_{e\infty}/\epsilon_0 \kappa T_e)^{1/2}} \tag{9-11}$$

If $T_e = T_i = T$, the effective sheath width is given by

$$X = \left[\frac{\epsilon_0 \kappa T}{n_{e\infty}|e|^2(Z_i + 1)}\right]^{1/2} \tag{9-12}$$

If $T_e \gg T_i$, the effective sheath width is given by

$$X = \left(\frac{\epsilon_0 \kappa T_i}{Z_i|e|^2 n_{e\infty}}\right)^{1/2} \tag{9-13}$$

The width of the sheath varies directly as the square root of the particle temperatures and inversely as the square root of the particle densities. Physically these results are quite plausible. It is the temperature or random thermal motion of the particles which supplies the energy necessary to create a charge unbalance in the first place. The greater the density of the charged particles, the lesser the distance necessary to create a given potential drop across the plasma.

It is appropriate now to examine the approximation of Eq. (9-8). In order to do this we must calculate the potential at the isolated material body V_0 as a function of temperature. For simplicity we shall assume that $T_e = T_i = T$. In equilibrium, the net electrical current to the wall must be zero. Thus at $x = 0$

$$\frac{n_{i0} Z_i |e| \overline{u_{i0}}}{4} - \frac{n_{e0}|e|\overline{u_{e0}}}{4} = 0 \tag{9-14}$$

where

$$n_{e0} = n_{e\infty} \exp\left(\frac{|e|V_0}{\kappa T}\right) \tag{9-15}$$

$$n_{i0} = n_{i\infty} \exp\left(-\frac{Z_i|e|V_0}{\kappa T}\right) \tag{9-16}$$

and $\overline{u_{i0}}$ and $\overline{u_{e0}}$ are calculated from Maxwell-Boltzmann statistics.

$$\overline{u_{i0}} = \left(\frac{8\kappa T}{\pi m_i}\right)^{1/2} \tag{9-17}$$

$$\overline{u_{e0}} = \left(\frac{8\kappa T}{\pi m_e}\right)^{1/2} \tag{9-18}$$

Substitution of Eqs. (9-15) to (9-18) into Eq. (9-14) yields

$$\frac{\exp(-Z_i|e|V_0/\kappa T)}{m_i^{1/2}} = \frac{\exp(|e|V_0/\kappa T)}{m_e^{1/2}} \tag{9-19}$$

where the condition of electrical neutrality at $x = \infty$ has again been invoked. Equation (9-19) can be solved by taking the logarithm of both sides of the equation and rearranging.

$$V_0 = -\frac{\kappa T}{2|e|(Z_i + 1)} \ln\left(\frac{m_i}{m_e}\right) \tag{9-20}$$

It is apparent from Eq. (9-20) that $|e|V_0$ is at best the same size as κT. Thus the approximation of Eq. (9-8) is not strictly valid. We can wriggle our way out of this predicament as follows: Our approximation [Eq. (9-8)] can be changed to read $|e|V(x) < \kappa T$, which, since the potential falls off very rapidly (exponentially) with x, is valid everywhere in the sheath except close to $x = 0$. Thus our solution is valid throughout most of the sheath, and since the potential must be continuous, it cannot be too far wrong in the region of $x = 0$.

As we expected from physical arguments, the potential at the isolated material body is negative with respect to the plasma. It should not have surprised us that the potential energy drop across the sheath and the thermal energy of the charged particles are of the same order of magnitude. It is the thermal energy of the plasma particles which sets up the sheath region. In equilibrium the thermal energy is balanced by the electric field which it induces.

9-3 The Debye Length

We now wish to find the potential variation around an extraneous charged particle, or, for that matter, around any charged particle, in the plasma. We shall set the origin of a spherical coordinate system at the site of the charged particle. Poisson's equation in spherical coordinates is

$$\frac{1}{r^2}\frac{\partial}{\partial r}\left[r^2 \frac{\partial V(r)}{\partial r}\right] = -\frac{\rho}{\epsilon_0} \tag{9-21}$$

where again the permittivity of the plasma is assumed to be that of free space. If the charged particles are assumed to be distributed in a Maxwell-Boltzmann fashion with $T_e = T_i = T$ and $|e|V(r) < \kappa T$, the charge density ρ can be written

$$\rho = -\frac{|e|^2 n_{e\infty}(Z_i + 1)}{\kappa T} V(r) \tag{9-22}$$

where $n_{e\infty}$ is the equilibrium electron density. For simplicity let us assume that $Z_i = 1$. The concurrent solution of Eqs. (9-21) and (9-22), with the boundary condition that there is a point charge at $r = 0$, yields

$$V(r) = \pm \frac{|e|e^{-r/\lambda_D}}{4\pi\epsilon_0 r} \tag{9-23}$$

where the Debye length λ_D is given by

$$\lambda_D = \left(\frac{\epsilon_0 \kappa T}{2n_{e\infty}|e|^2}\right)^{1/2} \tag{9-24}$$

The solution given in Eq. (9-23) can be verified by direct substitution. At small r the exponential term is approximately equal to unity, and the potential reduces to the potential of an unshielded point charge. As r becomes larger, the normal potential of an isolated point charge is reduced exponentially by the shielding of the charges of opposite sign. The spherical region in which there is a charge unbalance has an approximate radius of λ_D and is known as the "Debye sphere." Within the Debye sphere, thermal and electrical forces are in balance.

One of the outstanding characteristics of a plasma is its ability to maintain internal charge neutrality. The plasma can maintain internal charge neutrality only if the Debye length of the plasma is small compared with the physical size of the plasma. Thus a plasma can be defined as a collection of approximately equal numbers of mobile positive and negative charges for which λ_D, as defined by Eq. (9-24), is much less than the size of the plasma.

It is worthy of note that the concept of a Debye length was first introduced in connection with the behavioral description of strong electrolytes. A strong electrolyte is a liquid plasma.

PROBLEMS

9-1 How do the collision processes which may take place within a plasma enter into the calculation of the Debye length? Can collision interactions alter the derivation of the Debye length from that derivation we have given?

9-2 Find the Debye length in a plasma in which $q_i = Z_i|e|$, $q_e = -|e|$, and $T_e \neq T_i$.

9-3 Calculate the potential difference between a plasma and an isolated material object in contact with the plasma if the ion and electron temperatures are not equal. Discuss for $T_e \gg T_i$.

9-4 Plot the sheath thickness at a plasma-material body interface as a function of temperature for the case that $Z_i = 1$, $T_e = T_i = T$. Consider temperatures from 300 to 10^7°K and electron densities of
(*a*) 10^{22} electrons per cubic meter (the approximate density of an electron gas at 300°K, 1 mm Hg pressure)
(*b*) 10^{19} electrons per cubic meter (the approximate density of an electron gas at 300°K, 10^{-3} mm Hg pressure)
(*c*) 10^{16} electrons per cubic meter (the approximate density of an electron gas at 300°K, 10^{-6} mm Hg pressure)

9-5 How is the Debye length in a plasma related to the sheath thickness if $T_e = T_i = T$, $Z_i = 1$?

9-6 Plot the number of electrons in the Debye sphere as a function of temperature under the conditions given in Prob. 9-4.

9-7 In Chap. 3 we derived an expression for the random particle current density of a system of plasma particles under the condition that the particle density was independent of position coordinates. In Chap. 9 [Eq. (9-14)] we used this random current density in a situation in which the density is a function of position. Discuss the validity of this procedure.

9-8 If the potential of a conducting material object in contact with a plasma is varied with respect to the potential of the plasma, a current may flow through the plasma and through the material object. A plot of this current as a function of the potential difference between the material object and the

Fig. P9-8 Current versus potential difference between material object and plasma.

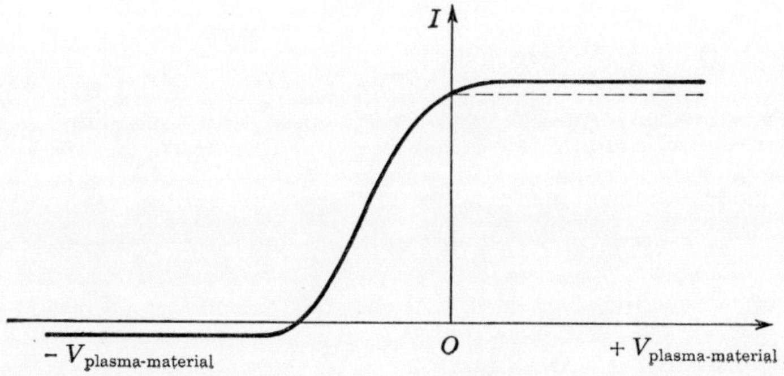

plasma is shown in Fig. P9-8. Explain qualitatively the current-voltage characteristics shown in Fig. P9-8. Notice the similarity between this characteristic and the I-V characteristic of a *p-n* junction or a rectifying metal-semiconductor junction.

BIBLIOGRAPHY

Debye, P., and W. Huckel: *Physikal. Z.*, **24**:183, 305 (1923).
On the Debye length in strong electrolytes.

Delcroix, J. L.: "Introduction to the Theory of Ionized Gases," Interscience Publishers, Inc., New York, 1960.
Sec. 10.1—On the relation between the Debye length and plasma sheaths.

Post, R. F.: Controlled Fusion Research: An Application of the Physics of High Temperature Plasmas, *Rev. Mod. Phys.*, **28**:338 (1956).
On the Debye length, with plots of the number of particles within the Debye sphere for various temperatures and particle densities.

chapter

10

Collisions and Radiation

10-1 Introduction

In the two theoretical approaches to the motion of plasma particles, collisions between the plasma particles have been considered. In the hydrodynamic approach, collisions are taken into account by the factor α or alternatively by ν_c. In the Boltzmann equation, collisions are taken into account by the collision term $(\partial f/\partial t)_{\text{collisions}}$. In neither theoretical approach is radiation which originates within the plasma considered. It is nevertheless a fact that if particles collide they must accelerate and that if charged particles accelerate they must radiate electromagnetic energy. Further, collisions may cause bound electrons to assume excited energy states within their parent atoms. When the electrons return to their original states, energy is released in the form of radiation. In many cases the radiation generated within a plasma is negligible and accordingly can be ignored. If radiative effects are important, the plasma equations must be amended to account for these effects.

In this chapter we shall examine the collision and radiative processes which may occur in a plasma. Several mathematical expressions concerning radiative effects will be admitted without proof. The derivations of these relations are outside the scope of this text.

10-2 Collisions

10-2-1 General. What is a collision? At first glance this question appears almost trivial. Nearly everyone, at one time or another, has played with marbles. Two marbles when shot together will touch or collide and then travel apart. The collision can be described by invoking simultaneously the concepts of conservation of energy and conservation of momentum. Still, the original question remains unanswered. Why do the marbles separate in the way that they do after "touching"? Why does one marble not pass through the other? Galaxies do. The answer to these questions is to be found in the word "touch." Truly, the marbles do not touch. They approach each other until the elementary particles in one marble exert a sufficient force on the elementary particles of the other marble to cause the marbles to move apart. So it is with all collisions. Elementary particles interact by virtue of gravitational, nuclear, and electromagnetic forces. The sum of the interactions is the collision, whether the colliding objects are electrons, marbles, pork chops, or galaxies. In classical collisions, energy and momentum are conserved. In nonclassical collisions, that is, quantum-mechanical and nuclear collisions, additional quantities may be conserved.

10-2-2 Elastic and Inelastic Collisions. Elastic collisions are those collisions in which kinetic energy is conserved; that is, the sum of the kinetic energies of the colliding particles is the same before as it is after the collision. Since charged particles must accelerate and therefore radiate energy when they suffer collisions, there is really no such thing as a perfectly elastic collision. The electromagnetic radiation emitted when two charged marbles collide, however, is too small to be measured. When electrons and ions collide elastically with energies below 500 ev, the radiation loss can, in almost all cases, be neglected.

Inelastic collisions are those collisions in which total energy is conserved, but kinetic energy is not. Some of the original kinetic energy of the colliding particles is absorbed internally by the particles. Thus when two pork chops collide, their deformation energy appears as heat. When an electron and a molecule collide, the molecule may carry off internal energy in the form of excited orbital electron states, molecular vibration, or molecular rotation. The molecule is capable of radiating this internal energy at various discrete wavelengths. Thus inelastic collisions are usually followed by the emission of radiation within the plasma.

10-2-3 Coulomb and Non-Coulomb Collisions. A Coulomb collision is a collision between two charged particles caused by the electrostatic force each particle exerts on the other. Coulomb collisions are long-range collisions; that is, two charged particles can "collide" while very far apart. The situation is similar to a comet "colliding" with the sun where the minimum distance of approach may be several million miles. Still, the comet and sun are both deflected from their original courses (the sun imperceptibly so, owing to its large mass). Because of the Debye shielding effect, the maximum distance for a long-range Coulomb collision within a plasma is limited to, approximately, a Debye length. Beyond this distance one charged particle does not feel the effect of the other. Because of their long-range characteristics, the majority of Coulomb collisions are small-angle collisions. The charged particles are deflected only slightly from their original directions of motion by the collisions.

We shall arbitrarily define a non-Coulomb collision as a collision in which one or both of the interacting particles are charge-neutral. If a plasma is not fully ionized, non-Coulomb collisions can be very important. In practice, the energy of the plasma particles must be on the order of thousands of electron volts to assure a fully ionized plasma. In many gas discharge tubes the degree of ionization is below 10 per cent. The electron energy in a cold-cathode discharge tube is usually less than 10 ev. For plasmas in which the degree of ionization is low, the electrons and ions collide primarily with neutral molecules. The form of these non-Coulomb collisions can be adequately described only by quantum mechanics. Suffice it to say that these non-Coulomb collisions are short-range collisions and therefore are similar in many respects to the collisions between marbles (elastic collisions) or between soft rubber balls (inelastic collisions). The collision between an electron or an ion and a neutral particle differs from the collision between two classical particles in that the apparent size of the quantum-mechanical particles is a function of the magnitude of the relative velocity of the two colliding particles.

For completeness we must also include under non-Coulomb collisions those interactions which take place because of the permanent or induced dipole moments of charge-neutral molecules. A dipole neutral can interact with a charged particle or with another dipole neutral. The origin of this type of non-Coulomb collision is certainly a Coulomb force. The fact that one or both of the colliding particles are charge-neutral excludes this type of interaction from the arbitrarily defined Coulomb collision.

It is interesting to note that all of the collisions which we have called non-Coulomb (we exclude nuclear and gravitational collisions) have their origins in electrical interactions. Although the interactions are electrical, the characteristics of the interactions can be explained only by invoking quantum-mechanical considerations.

10-2-4 Collision Cross Section, Mean Free Path, and Collision Frequency. We shall discuss the collision characteristics of two interacting types of plasma particles. We first assume that the two types of plasma particles (which we shall call "type 1" and "type 2") behave as if they were hard elastic spheres of radii r_1 and r_2, respectively. Later we shall alter this assumption to account for more realistic types of interactions.

In order that two hard elastic spheres of radii r_1 and r_2 collide, their centers must be a distance $(r_1 + r_2)$ apart. Let us assume that the type-2 particles are stationary and of uniform density and that a given type-1 particle moves with a speed u. We can simplify the problem by representing the moving particle as a point and the stationary particles as spheres of radii $(r_1 + r_2)$. The criterion for a collision remains unchanged.

Suppose we follow the motion of the point as it travels a distance dx. The probability that the type-1 particle will make a collision with a type-2 particle in going dx is $\pi(r_1 + r_2)^2 n_2\, dx$, where n_2 is the uniform density of the stationary particles. An examination of Fig. 10-1 will make the expression for the collision probability clearer. The probability that a moving point will collide with a stationary sphere of radii $(r_1 + r_2)$ in going dx is the quotient of the total cross-sectional area of the spheres which the point sees in a given area A perpendicular to the motion of the point divided by that area. The total cross-sectional area of the spheres is the total area of their projection on A, as shown in Fig. 10-1. The projected area of one sphere is $\pi(r_1 + r_2)^2$. The number of spheres in the volume $A\, dx$ is $n_2 A\, dx$. We have implicitly assumed, in this derivation of the collision probability, that there are no overlapping spherical projections on A. An equivalent assumption would be that the interparticle distance is very large

Fig. 10-1 Diagram from which collision probability may be deduced.

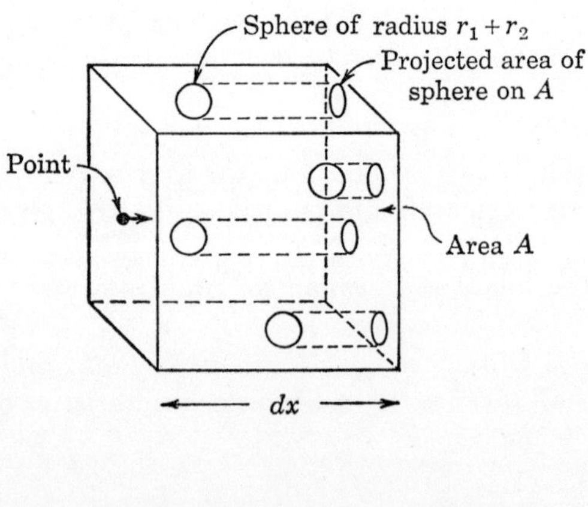

compared with $(r_1 + r_2)$. The effective diameter of an atom is approximately 10^{-10} m. The spacing between plasma particles at atmospheric pressure and room temperature is about 10^{-8} m.

Suppose we propel J_{01} points per unit area per second of speed u into the uniform density of stationary spheres. We shall assume that if a particle makes a collision, it is removed from the original particle current. If $J_1(x)$ is the particle current density of the points a distance x into the spheres, the change in J_1 in the next interval dx represents the number of points which will make collisions in dx after having gone x. This number, dJ_1, is found by multiplying the probability of any one point making a collision in dx times the current density of points at x.

$$dJ_1 = -J_1\pi(r_1 + r_2)^2 n_2 \, dx \tag{10-1}$$

The solution to Eq. (10-1) subject to the given initial conditions is

$$J_1 = J_{01} \exp\left[-n_2(r_1 + r_2)^2 \pi x\right] \tag{10-2}$$

To find the average distance any one point travels before making a collision, that is, the *mean free path* of the type-1 particles, we must add each free path weighted by the current density of points that have that particular free path and divide by the original current density J_{01}.

$$\lambda_1 = \frac{\int_{J_{01}}^{0} x \, dJ_1}{J_{01}} \tag{10-3}$$

Equations (10-1) to (10-3) can be combined to yield

$$\lambda_1 = \int_0^\infty x n_2 \pi (r_1 + r_2)^2 \exp\left[-n_2\pi(r_1 + r_2)^2 x\right] dx \tag{10-4}$$

where the minus sign in Eq. (10-1) has been ignored. After evaluating the integral in Eq. (10-4), we find that

$$\lambda_1 = \frac{1}{n_2\pi(r_1 + r_2)^2} \tag{10-5}$$

The mean free path of the type-1 particles depends only on the density of the type-2 particles and on the effective cross-sectional area of the stationary spheres. We shall define $\pi(r_1 + r_2)^2$ as the effective collision "cross section" σ_{12} for a collision between type-1 and type-2 particles. It follows that

$$\lambda_1 = \frac{1}{n_2\sigma_{12}} \tag{10-6}$$

We have derived Eq. (10-6) for the case that the type-2 particles remain stationary. Equation (10-6) is, however, a good approximation as long as the speed of the type-1 particles greatly exceeds that of the type-2 particles. If the motion of the type-2 particles is comparable with that of the type-1 particles, it is physically apparent that the mean free path of the type-1 particles will be decreased from the value given by Eq. (10-6). Under these circumstances it is necessary to rederive the expression for the collision probability of a type-1 particle taking into account the motion of the type-2 particles.

Equation (10-6) can be used as a definition for the effective collision cross section for collisions between particles which do not behave like hard elastic spheres. The mean free path can be determined experimentally by observing the exponential decay of a beam of one type of particles fired into an approximately stationary group of another type of particles, $J_1 = J_{01}e^{-x/\lambda_1}$. The effective collision cross section is calculated from Eq. (10-6). The cross section may also be determined theoretically. It is found that the cross sections for interactions between actual plasma particles are rather complex functions of the magnitude of the relative velocity of the two particles which are colliding. Only in rare cases is the collision cross section constant, as in the interaction of two hard elastic spheres. The collision cross section for Coulomb collisions within a plasma varies inversely as the fourth power of the magnitude of the relative particle velocity. The total collision cross sections for several non-Coulomb collisions are plotted in Fig. 10-2. The non-Coulomb collisions considered are between electrons and neutral molecules. The cross sections are plotted as a function of the energy associated with the relative velocity between the electron and the molecule. The cross sections shown in Fig. 10-2 are *total* collision cross sections. They represent processes by which particles are removed from the original particle beam. The total collision cross section can be decomposed into a number of component collision cross sections. For instance, there is an effective collision cross section for the excitation of any given bound electron state during a collision between a molecule and a free electron. There is an effective cross section for the attachment of an electron to a molecule during a collision. Any total collision cross section of the type shown in Fig. 10-2 is the sum of all component cross sections, each of which represents a process by which a particle can be scattered from its original direction of motion. It is also possible to define

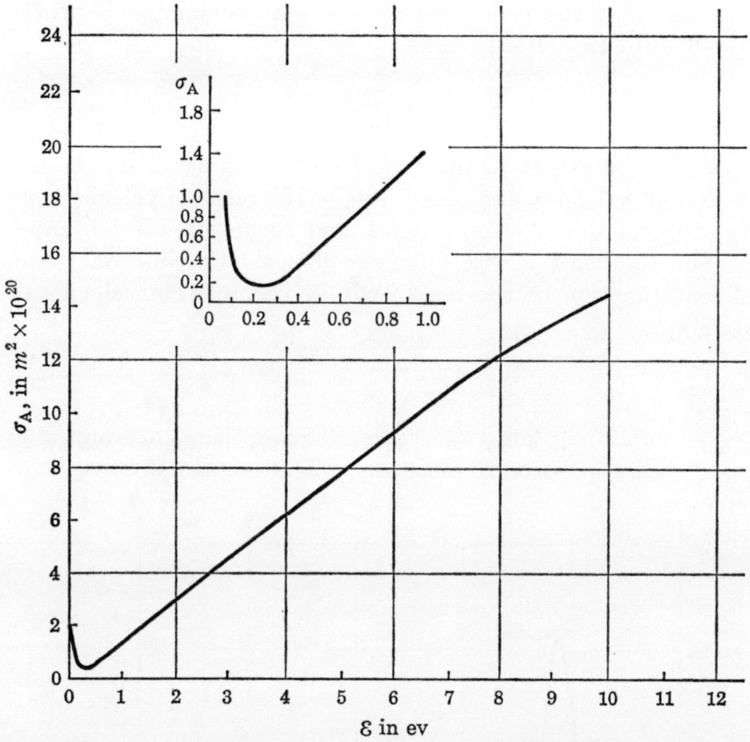

Fig. 10-2A Cross section for an electron-argon collision. [*Data adapted from D. Barbiere, Phys. Rev.*, **84**: 653 (1951); *B. Kivel, Phys. Rev.*, **116**: 926 (1959); *L. S. Frost, Bull. Am. Phys. Soc., Paper U*-9, *June* 15, 1960.]

cross sections for the production of new plasma particles, that is, ionization cross sections. In fact, it is possible to define as many collision cross sections as there are types of collision processes.

The mean free path of a type-1 particle of speed u is the average distance that the type-1 particle travels between collisions. The collision frequency of the particle is found by dividing the speed of the particle by its mean free path.

$$\nu_1 = \frac{u}{\lambda_1} \tag{10-7}$$

where we have used the notation ν_1 to distinguish the collision frequency of the particle with speed u from the effective collision frequency ν_c defined in connection with the momentum transfer equations.

In the case that the type-2 particles are stationary, substitution of Eq. (10-6) into Eq. (10-7) yields

$$\nu_1 = n_2\sigma_{12}u \tag{10-8}$$

The collision frequency of the type-1 particle depends on the density of the type-2 particles, the magnitude of the relative velocity between the type-1 and the type-2 particles, and the total collision cross section. For the hard elastic sphere collision, the collision frequency is a linear function of the magnitude of the relative velocity of the colliding spheres.

Fig. 10-2B Cross section for an electron–molecular nitrogen collision. [*Data adapted from R. B. Brode, Rev. Mod. Phys.*, **5**: 257 (1933).]

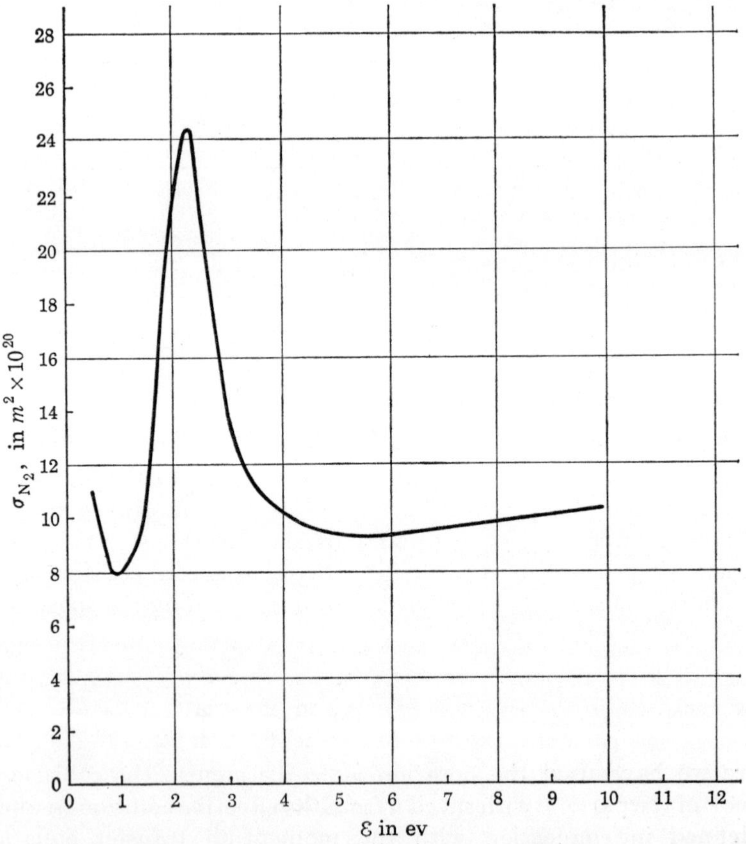

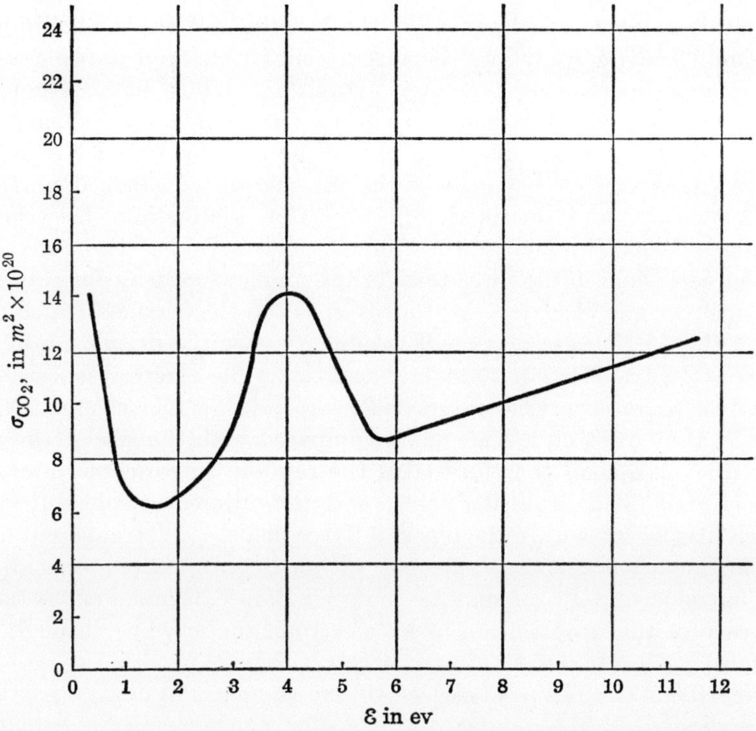

Fig. 10-2C Cross section for an electron–carbon dioxide collision. [*Data adapted from R. B. Brode, Rev. Mod. Phys.*, **5**: 257 (1933)].

Equation (10-8) can be generalized to describe any type of collision if σ is the appropriate collision cross section. Thus there is a collision frequency for electron excitation just as there is a cross section for electron excitation. In general, both the collision frequency and the collision cross section are functions of the magnitude of the relative velocity of the colliding particles.

10-2-5 Collisions and the Equations of Motion: The Momentum Transfer Equation. Now that we have classified and discussed the types of collisions that can occur within a plasma, we are prepared to consider how these collisions can be included in the mathematics of plasma physics. In general, this is a very sticky business, in which many assumptions and approximations are made. We shall look first at how the effects of collisions are included in the momentum transfer equations.

At this point we must again ask the question "What is a collision?" Coulomb collisions involve the interaction of charged particles with the electric fields of other charged particles. Thus, in undergoing a Coulomb collision, a charged particle is acted upon by an electric force. There is, however, a force term in the momentum transfer equation which takes explicit account of the interaction of charged particles with electric and magnetic fields. The distinction between a "collision" electric field and a "force" electric field is the following: The electric field in the force term of the momentum transfer equation is a macroscopic electric field; that is, it is the electric field which acts on all the particles in the small volume d^3r (see the derivation of the momentum transfer equation in Chap. 4). The electric fields which give rise to interparticle interactions are microscopic electric fields; that is, they exist on a scale small compared with the size of the volume d^3r. We shall therefore label the random microscopic interparticle electric fields "collision fields" and any ordered or coherent fields which may occur within the plasma "force fields." The coherent electric fields may occur spontaneously within the plasma (as we shall see in Chaps. 11 and 12) or may be imposed from external sources (as in the case of the propagation of an electromagnetic wave through the plasma).

The effects of collisions between electrons, ions, and neutral particles are included in the momentum transfer equations by postulating that the electron, ion, and neutral-particle fluids exert frictional drag forces on one another. If the plasma does not contain neutral particles, the magnitude of the frictional-force term for the electron fluid will be equal to the magnitude of the frictional-force term for the ion fluid. If neutral particles are present within the plasma, the frictional terms for the electron and ion fluids will no longer be equal since the ion-neutral interaction differs from the electron-neutral interaction. The effects of electron-electron, ion-ion, and neutral-neutral interactions are included in the momentum transfer equations by postulating an electron pressure, an ion pressure, and a neutral-particle pressure.

We have noted that the term $(\alpha_{ei}n_i + \alpha_{en}n_n)/m_e$ in the momentum transfer equation for electrons has the dimensions of frequency, and therefore we have defined this term as the electron *effective* collision frequency. In a similar manner we have defined an ion *effective* collision frequency and an neutral-particle *effective* collision frequency. The defined effective collision frequencies are obviously in some way related to the actual collision frequencies of an average plasma electron, ion, or neutral particle since collisions are the cause of the damping or frictional forces. In order to determine the exact relationship between the effective collision frequencies defined in the momentum transfer equations and the actual collision frequencies within a plasma, we must derive the momentum transfer equations from the appropriate Boltzmann equations (see Appendix A). The results of this derivation *for the electrons* often indicate that the defined effective collision frequency is of the same order of magnitude as the actual collision frequency (electron-ion and electron-neutral) of an average plasma electron. We can get a physical feeling for this result by examining the electron momentum transfer equation. If we multiply Eq. (4-10) by the mass of the electron, each term in the resultant equation represents a rate of change of electron average momentum. The rate of change of electron average momentum due to collisions is therefore $m_e\mathbf{v}_e\nu_{ce}$. In one second the electron average momentum is lost ν_{ce} times. If ν_{ce} is to correspond to the actual collision frequency of an average plasma electron, that electron must lose its average momentum at each collision. When a ping-pong ball collides with a basketball, the momentum of the ping-pong ball is radically altered, but the momentum of the basketball is not changed appreciably. It is therefore reasonable that an electron, in colliding with ions and with neutral particles, will lose (within an order of magnitude) its average momentum at each collision. The defined effective collision frequency may therefore be of the same order of magnitude as the actual collision frequency of an average plasma electron.

The effective collision frequencies for the ions and the neutral particles may differ appreciably from the appropriate actual collision frequencies. The reason for this is that the ions and neutrals may not lose their average momenta at each collision. In fact, in collisions with the lighter electrons, the ions and neutrals will lose very little momenta. Thus in a highly ionized plasma (predominantly electron-ion collisions), the ion effective collision frequency must be much smaller than the actual collision frequency of an average plasma ion. In a weakly ionized plasma (predominantly ion-neutral, electron-neutral collisions), however, the ion effective collision frequency may be a good approximation to the actual collision frequency of an average plasma ion since particles of approximately equal mass (ions and neutrals) can exchange significant portions of their momenta in a collision.

We may conclude that it is sometimes a reasonable approximation to use the actual collision frequencies of the appropriate average plasma particles in the momentum transfer equations. The particle collision frequencies or, alternatively, the collision cross sections for the various collision interactions may be determined from theory or by separate experiment. Note that both the effective and the actual collision frequencies are functions of particle density, which may in turn be a function of position coordinates.

Often the plasma particles will have large random velocities. If the particular plasma phenomenon under study requires that a small nonrandom velocity (for instance, the velocity induced by the electric field of an electromagnetic wave) be considered, it is usually a reasonable approximation to treat the effective collision frequency in the momentum transfer equation as independent of the average velocity of the particles. The change in the collision frequency caused by the small coherent velocity is, in these cases, assumed to be negligible compared with the collision frequency attributable to the random velocity.

10-2-6 Collisions and the Equations of Motion: The Boltzmann Equation. The effects of all relevant interparticle interactions must be included in the collision term of the Boltzmann equation. Coulomb collisions involve the interactions of particles with electric fields. In addition to the collision term in the Boltzmann equation there is an electric field term. The electric field appearing explicitly in the Boltzmann equation takes account of the drift of particles in velocity coordinates due to electric forces. Clearly a Coulomb collision involves a drift of particles in velocity coordinates due to electric forces. The distinction between a "drift" electric field and a "collision" electric field is the following: The electric field that appears explicitly in the Boltzmann equation is a macroscopic electric field which acts on all particles whose representative points are within the phase-space volume $d\mathcal{V}$ (see the derivation of the Boltzmann equation in Chap. 4). The electric fields that give rise to interparticle interactions are microscopic fields which exist on a scale small compared with the volume d^3r occupied by particles whose representative points are in $d\mathcal{V}$. We shall therefore label the random microscopic rapidly varying electric field–particle interactions "collisions," and any ordered electric field which may occur, the "drift" field.

In deriving the Boltzmann equation, we included in the collision term only those collision processes which cause a plasma particle to exhibit a spontaneous velocity change without a position coordinate change. Clearly, Coulomb collisions do involve a position coordinate change and thus can be included in the collision term of the Boltzmann equation only in an approximate form.

Once it has been decided what interactions are to be included in the collision term, the inclusion of these interactions is in principle straightforward. It is another matter, however, to formulate the collision term in such a manner that the Boltzmann equation can be solved. The procedure generally involves a number of expansions and approximations.

We shall formulate the collision term for a special case: the collision of type-1 particles with stationary type-2 particles of uniform density. We can write the collision term as follows:

$$\left(\frac{\partial f}{\partial t}\right)_{\text{collisions}} = \left(\frac{dN_{\text{in}} - dN_{\text{out}}}{d\mathcal{V}\, dt}\right)_{\text{collisions}} \tag{10-9}$$

where f is the distribution function for the type-1 particles, dN_{in} is the number of phase-space points representing type-1 particles which are scattered into the phase-space volume $d\mathcal{V} = du_x\, du_y\, du_z\, dx\, dy\, dz$ by collisions in time dt, and dN_{out} is the number of representative points scattered out of $d\mathcal{V}$ by collisions in time dt. We shall first calculate dN_{out}. There are $f(u_x,u_y,u_z,x,y,z,t)\, d\mathcal{V}$ representative points in $d\mathcal{V}$. The number of representative points scattered out of $d\mathcal{V}$ in time dt is that number $f\, d\mathcal{V}$ multiplied by the collision frequency of a particle with speed u and by dt.

$$dN_{\text{out}} = f(u_x,u_y,u_z,x,y,z,t)\,\nu_1(u)\, du_x\, du_y\, du_z\, dx\, dy\, dz\, dt \tag{10-10}$$

where $u = (u_x^2 + u_y^2 + u_z^2)^{1/2}$. After making use of the expression for the collision frequency from Eq. (10-8), we find that

$$dN_{\text{out}} = n_2\sigma_{12}(u)uf(u_x,u_y,u_z,x,y,z,t)\, d\mathcal{V}\, dt \tag{10-11}$$

The calculation of the scattering of representative points into $d\mathcal{V}$ in time dt is more complex. Representative points with many velocities can be scattered into $d\mathcal{V}$ by the appropriate type of collision. We shall designate the speed u' as the speed associated with a representative point prior to its being scattered into $d\mathcal{V}$. Thus the effect of the collision is to change u' to u. If a representative point originally in $d\mathcal{V}'$ is to be scattered into $d\mathcal{V}$, the plasma particle which the point represents must be scattered a definite angle, which we shall call Ω', from its original direction of motion. We shall define $\Sigma_{12}(u',\Omega')\, d\Omega'$ as the cross section for the scattering of a plasma particle of original speed u' in the direction Ω', into the solid angle $d\Omega'$, such that the representative point for that particle may end up in $d\mathcal{V}$. The number of representative points in $d\mathcal{V}'$ is

$$f(u_x',u_y',u_z',x,y,z,t)\, du_x'\, du_y'\, du_z'\, dx\, dy\, dz$$

where we recall from Chap. 4 that collisions do not change the location of the representative points in position coordinates. Thus the number of representative points scattered into $d\mathcal{V}$ from one $d\mathcal{V}'$ is

$$n_2\Sigma_{12}(u',\Omega')\, d\Omega'\, u'f(u_x',u_y',u_z',x,y,z,t)\, d\mathcal{V}'\, dt \tag{10-12}$$

There are, however, a number of volumes $d\mathcal{V}'$ from which representative points can be scattered into $d\mathcal{V}$. For each $d\mathcal{V}'$ there is a specific value of Ω'. The total number of representative points scattered into $d\mathcal{V}$ from all possible $d\mathcal{V}'$ can be found by integrating over all Ω', remembering that Ω' is functionally related to u' and that for each Ω' there will be an appropriate u'.

$$dN_{\text{in}} = \iint_{\Omega'} n_2\Sigma_{12}(u',\Omega')u'f(u_x',u_y',u_z',x,y,z,t)\, d\mathcal{V}'\, dt\, d\Omega' \tag{10-13}$$

We can combine Eqs. (10-9), (10-11), and (10-13) to give the total collision term

$$\left(\frac{\partial f}{\partial t}\right)_{\text{collisions}} = -n_2\sigma_{12}(u)uf(u_x,u_y,u_z,x,y,z,t) + \iint_{\Omega'} \frac{n_2\Sigma_{12}(u',\Omega')u'f(u_x',u_y',u_z',x,y,z,t)\, du_x'\, du_y'\, du_z'\, d\Omega'}{du_x\, du_y\, du_z} \tag{10-14}$$

In order to simplify the collision term further, it is necessary to know the relationship between u, u', and Ω'. Functional relationships can be written for both elastic and inelastic collisions. The primed velocity quantities in the collision integral can therefore be eliminated. The Boltzmann equation becomes an integrodifferential equation since it contains both integrals and derivatives of the distribution function. Several techniques have been developed for the solution of this form of the Boltzmann equation.

10-3 Radiation

10-3-1 General. Radiation originating from the plasma particles cannot be included in the mathematics of plasma physics without making those equations very difficult to solve. It is primarily important to know whether radiation effects are significant in relation to general plasma behavior. If they are unimportant, the plasma equations we have derived are perfectly adequate. If internal radiation is significant, this information must be somehow included in the theoretical plasma analysis. In high-energy low-density plasmas, the internal plasma radiation escaping from the plasma results in a reduced energy for the plasma particles from the value predicted without taking account of the radiation. In high-energy high-density plasmas (stars, for example), the radiation pressure due to internal radiative emission cannot be ignored. For instance, if a plasma is in thermal equilibrium with black-body radiation at 10 kev, the radiation pressure would amount to about 10^{11} atm. In addition, the energy leaving a high-density plasma by black-body radiation is very sizable.

We shall discuss the following types of radiation which occur in plasmas: (1) excitation and recombination radiation, (2) bremsstrahlung, (3) cyclotron radiation, and (4) black-body radiation.

10-3-2 Excitation and Recombination Radiation. Excitation radiation is, in general, the greatest source of internal plasma radiation in low-density non-fully ionized plasmas with particle energies below about 1,000 ev. This is a very fortunate state of affairs for the manufacturers of neon lights.

The principal characteristic of excitation radiation is that the radiation is emitted at discrete frequencies. Whenever a plasma contains atoms whose orbital electrons have not been completely stripped, this discrete or line radiation will appear.

In general, the higher the atomic number of the atom or molecule, the higher the energy that a plasma must have in order completely to strip all ions of their orbital electrons. At plasma energies of 50 ev, a hydrogen plasma is essentially completely ionized. At 2,000 ev, about 1 per cent of the ions in an oxygen plasma are not stripped of all of their electrons. If an atom retains orbital electrons, these electrons can be excited to higher-energy states by particle impact. Upon returning to their original states, the orbital electrons give up energy in the form of radiation at various discrete frequencies.

Free electrons in a plasma can recombine with ions or, in some cases, can be captured by neutral molecules. The energy lost by the electrons in these processes may appear as radiation. Since the free electrons have a velocity distribution, there will be a spectrum of energy losses and hence a spectrum or continuum of radiation emitted. Each recombination continuum, corresponding to each allowed energy level that an electron can fall into, will have a low-frequency limit representing the capture of electrons with very low kinetic energies.

10-3-3 Bremsstrahlung. In the course of a Coulomb collision, both charged particles participating in the collision are accelerated. Accelerating charges radiate electromagnetic energy in proportion to the square of their acceleration. The radiation which arises from Coulomb collisions is called "bremsstrahlung" or "braking radiation." At energies below 50,000 ev, plasma bremsstrahlung arises almost entirely from electron radiation in electron-ion interactions. The ions are too massive to suffer the necessary acceleration. Since an electron is free before and after the Coulomb collision, bremsstrahlung is sometimes referred to as "free-free" radiation. Bremsstrahlung is continuous radiation. All frequencies are present. For 10-ev electrons, the wavelength at which the maximum energy is radiated by bremsstrahlung is about 1,000 A. For 1,000-ev electrons, the wavelength at which the maximum energy is radiated by bremsstrahlung is about 10 A; for 100,000-ev electrons, about 0.1 A. For electron energies below the hundreds of electron volts range, bremsstrahlung can usually be neglected. At these low electron energies it is necessary to have relatively high electron and ion densities in order that the total bremsstrahlung radiated by the plasma electrons be appreciable. Bremsstrahlung, therefore, usually occurs in the X-ray to ultraviolet wavelength range. The approximate power radiated per unit volume by bremsstrahlung in a plasma is given by

$$P_B \cong 10^{-38} Z_i^2 n_i n_e \bar{\mathcal{E}}_e^{1/2} \qquad \text{watts/m}^3 \tag{10-15}$$

where all quantities are in mks units except the electron average energy, which is expressed in electron volts.

10-3-4 Cyclotron Radiation. If there is a magnetic field present within a plasma, the plasma particles will, in the absence of collisions, orbit in a circular fashion about the magnetic field lines. Uniform circular motion requires a constant center-directed acceleration. Thus charge particles orbiting in a magnetic field must radiate. Cyclotron radiation, like bremsstrahlung, is continuous radiation. The frequency spectrum of cyclotron radiation, however, has peaks at the cyclotron frequency and at the first few harmonics of the cyclotron frequency. Cyclotron radiation further differs from bremsstrahlung in that the wavelength for maximum power emission in man-made plasmas is usually much longer, from infrared to microwave, for cyclotron radiation. An approximate expression for the power radiated per unit volume in the cyclotron mode is given by

$$P_C \cong 10^{-19} n_e \bar{\mathcal{E}}_e |\mathbf{B}|^2 \qquad \text{watts/m}^3 \tag{10-16}$$

where all quantities are in mks units except the electron average energy, which is expressed in electron volts. The cyclotron radiation loss from a plasma increases more rapidly with increasing electron energy than the bremsstrahlung loss. The ratio of the cyclotron radiation power to the bremsstrahlung power can be found by combining Eqs. (10-15) and (10-16).

$$\frac{P_C}{P_B} \cong 10^{19} \frac{|\mathbf{B}|^2 \bar{\mathcal{E}}_e^{1/2}}{Z_i^2 n_i} \tag{10-17}$$

Cyclotron (or, as it is sometimes called, synchrotron) radiation received its name from a similar radiation which occurs in cyclotrons and synchrotrons. The radiation losses in these particle-accelerating machines make it necessary to adopt sophisticated techniques in order to achieve efficient operation.

10-3-5 Black-body Radiation. If the electromagnetic radiation generated within a plasma by the three mechanisms we have discussed is absorbed and remitted many times before reaching the boundaries of the plasma, the radiation will come into equilibrium with itself and with the plasma particles. This equilibrium radiation is called "black-body" radiation since it is identical in form with the radiation emanating from a solid-state black body whose temperature is equal to that of the plasma. The total radiation flux from the surface of a black body is dependent only on the temperature of the body

$$I = \zeta T^4 \tag{10-18}$$

where $\zeta = 5.7 \times 10^{-8}$ joule/(°K)^4(m^2-sec). Equation (10-18) is called the "Stefan-Boltzmann law."

The plasma can be opaque to certain wavelengths and transparent to others. In this case the radiation emanating from the plasma will be a combination of a quasi-black-body radiation and one or more of the three original radiative forms.

10-3-6 The Observation of Radiation as a Diagnostic Technique. The field of astronomy is concerned to a great extent with the diagnosis of plasma radiation. A study of the radiation emitted from man-made plasmas yields a great deal of information about the plasmas.

From a knowledge of the relative intensities of two spectral lines, an estimate of the average electron energy can be made. From a knowledge of the absolute intensities of the spectral lines, the atom and ion number densities can be calculated. No spectral line is really a discrete line. The spectral lines are broadened appreciably by a number of effects. A measurement of the line broadening yields information about the plasma. Because the atoms and ions of a plasma have a velocity distribution, the Doppler effect will cause a range of frequencies to be emitted on either side of any discrete emission line. An atom moving away from an observer appears to radiate at a lower frequency than an atom moving toward an observer. The Doppler effect thus broadens the spectral lines. The Doppler broadening is directly related to the average energy of the radiating ions or atoms. The average ion or atom energy can therefore be determined by a measurement of the Doppler broadening. The Stark effect, the splitting of spectral lines due to an electric field, also causes a broadening of the spectral lines. The electric fields which induce the Stark effect are the random microscopic fields between the charged particles. Thus the charged-particle density can be determined by measuring the Stark effect. The effect of interparticle collisions is also to broaden the spectral lines. Collisions disturb the phase of the radiative emission, and this phase disturbance results in a line broadening. Collision frequencies may be determined from a measurement of the collision broadening. In general, all of the effects which cause the spectral lines to be broadened occur simultaneously. In order to obtain useful information, it is necessary to find situations in which one of the several broadening effects is dominant.

From an analysis of the continuum spectrum due to bremsstrahlung, the number density and the average energy of the plasma electrons can be determined. From an analysis of the continuum spectrum due to black-body radiation, the black-body temperature of a plasma can be calculated. A plasma is opaque to radiation whose frequency is near the electron plasma frequency. Thus for a range of frequencies about the electron plasma frequency, a plasma will emit black-body radiation. Since the electron plasma frequency is usually in the microwave frequency range, microwave techniques are used to measure the black-body temperature of a plasma. The cyclotron radiation emitted from a plasma can be analyzed to determine properties of the magnetic field which accelerates the charged plasma particles. If the form of magnetic field is known, the cyclotron radiation data can be used to find the number density and the average energy of the plasma electrons. The three types of continuum radiation we have just discussed occur in separated frequency ranges so that there is, in general, no difficulty in measuring all three types of radiation simultaneously. It should be noted that in order to analyze theoretically bremsstrahlung and cyclotron radiation, an electron phase-space distribution function must be assumed. Since Maxwell-Boltzmann statistics is usually used, the mathematical analyses may not accurately describe the physical situation. There may therefore be inaccuracies in the values of the electron density and electron average energy determined by comparing the theory with the experimental data.

By using a combination of appropriate spectroscopic techniques and theoretical analyses, the average values of various plasma parameters can be determined. It is very difficult to determine electron and ion distribution functions by spectroscopic means.

PROBLEMS

10-1 Find the mean free path of an electron within a binary gas mixture of uniform density. The two types of gas molecules have densities n_A and n_B, respectively, and cross sections for electron interactions σ_A and σ_B, respectively. Show that

$$\frac{1}{\lambda} = \frac{1}{\lambda_A} + \frac{1}{\lambda_B}$$

where

$$\lambda_A = \frac{1}{n_A \sigma_A} \qquad \lambda_B = \frac{1}{n_B \sigma_B}$$

10-2 Under what conditions is the effective collision frequency for the neutral plasma particles of the same order of magnitude as the actual collision frequency of an average neutral particle?

10-3 Plot the approximate electron-neutral collision frequency in a plasma as a function of neutral-particle density for n_n between 10^{16} particles per cubic meter (about 10^{-6} mm Hg at 300°K) and 10^{22} particles per cubic meter (about 1 mm Hg at 300°K) if $\overline{\mathcal{E}_n} = 0.03$ ev and

(*a*) $\overline{\mathcal{E}_e} = 1$ ev, $\sigma_{en} = 10^{-19}$ m²

(*b*) $\overline{\mathcal{E}_e} = 100$ ev, $\sigma_{en} = 10^{-20}$ m²

10-4 Prove that in an elastic collision between a moving particle of mass m and initial energy $\mathcal{E}_{0m}$ and an initially stationary particle of mass M, the fractional energy loss of the particle of mass m is

$$\frac{\mathcal{E}_{0m} - \mathcal{E}_m}{\mathcal{E}_{0m}} = \frac{4mM}{(m + M)^2} \cos^2 \theta$$

and, if $m \ll M$, the fractional energy loss is

$$\frac{\mathcal{E}_{0m} - \mathcal{E}_m}{\mathcal{E}_{0m}} \cong \frac{2m}{M} (1 - \cos \psi)$$

A diagram of the interaction is shown in Fig. P10-4.

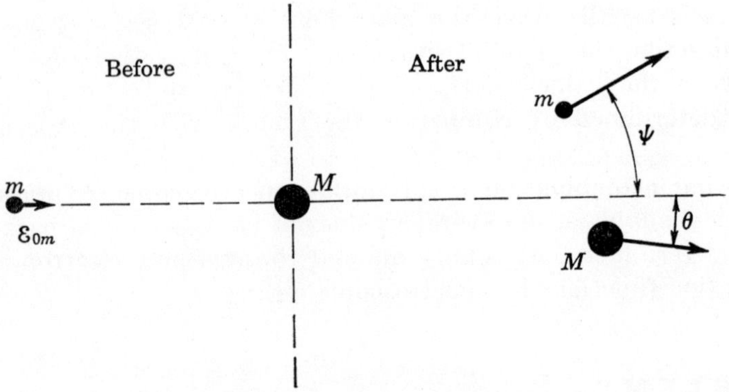

Fig. P10-4 The collision interaction.

10-5 Consider a plasma of volume 10^{-3} m³ contained in a magnetic field whose flux density is 5 webers/m². What values of electron average energy and density will produce 1 watt of bremsstrahlung and 1 watt of cyclotron radiation? For the specified radiation emission, plot the electron density as a function of the electron energy for electron energies between 1 and 10^4 ev.

BIBLIOGRAPHY

Allis, W. P.: Motion of Ions and Electrons, "Handbuch der Physik," vol. 21, Springer-Verlag OHG, Berlin, 1956 (in English).
p. 404—On the solution of the Boltzmann equation.
p. 488—On the formulation of $(\partial f/\partial t)_{\text{collisions}}$.

Brown, S. C.: "Basic Data of Plasma Physics," John Wiley & Sons, Inc., New York, 1959.
Contains experimental data on the various collision processes.

Donohoe, R. J., and R. F. Majkowski: Spectroscopic Measurements of Temperatures and Densities in a Cesium Plasma, *J. Appl. Phys.*, **33**:3 (1962).

Glasstone, S., and R. H. Lovberg: "Controlled Thermonuclear Reactions," D. Van Nostrand Company, Inc., Princeton, N.J., 1960.
p. 25—A general discussion of plasma radiation.
p. 84—A discussion of Coulomb interactions.

Heitler, W.: "The Quantum Theory of Radiation," Oxford University Press, London, 1944.
Both classical and quantum-mechanical discussions of all types of radiation processes.

Kennard, E. H.: "Kinetic Theory of Gases," McGraw-Hill Book Company, Inc., New York, 1938.
On the calculation of collision frequencies, cross sections, and mean free paths.

Lochte-Holtgreven, W.: Production and Measurement of High Temperatures, *Repts. Progr. Phys.*, **21**:312 (1958).
A review with references of the types of plasma radiation and of the optical techniques used to determine the properties of a plasma.

Menzel, D. H. (ed.): "Selected Papers of Physical Processes in Ionized Plasmas," Dover Publications, Inc., New York, 1962.
On stellar spectroscopy.

Molmud, P.: Langevin Equation and the AC Conductivity of Non-Maxwellian Plasmas, *Phys. Rev.*, **114**:29 (1959).
On the relation between the actual electron collision frequency in a plasma and the defined *effective* collision frequency.

Morse, P. M., W. P. Allis, and E. S. Lamar: Velocity Distributions for Elastically Colliding Electrons, *Phys. Rev.*, **48**:412 (1935).
On the formulation of $(\partial f/\partial t)_{\text{collisions}}$ for elastic marblelike collisions.

Rose, D. J., and M. Clark, Jr.: "Plasmas and Controlled Fusion," John Wiley & Sons, Inc., New York, 1961.
chap. 2— Collision Phenomena: Basic Theory and Fusion Cross Sections.
chap. 8—Coulomb Interactions and Their Consequences.
chap. 11—Radiation Losses from a Plasma.

Sengupta, D. L.: The Electrical Conductivity of a Partially Ionized Gas, *Proc. IRE*, **49**:1872 (1961).
On the solution of the Boltzmann equation considering both electron-neutral collisions and electron-ion collisions.

chapter

11

Longitudinal Oscillations of Plasma Electrons

11-1 Introduction

One of the characteristics of the plasma state is the ability of the plasma to maintain internal charge neutrality. We have seen an example of this behavior in our discussion of the Debye length: A microscopic unbalance in charge due to the random thermal motion of the plasma particles is opposed by an induced electric field which tends to restore charge neutrality. We shall consider in this chapter another result of the attempt of the plasma to maintain internal charge neutrality—the longitudinal oscillations of plasma electrons. The electron oscillations to be discussed are called "longitudinal" because the average motion of the electrons and the electric field which causes that motion are both parallel to the direction of propagation of the oscillations. We shall find that there is no magnetic field associated with the longitudinal oscillations.

We shall first consider that the plasma particles exhibit no random thermal motion, that is, that the plasma is at zero temperature. Each charged particle will then assume an equilibrium position such that the total force on the particle from all other particles is zero. The plasma is as "charge-neutral" as is possible. Now let us remove one electron a small distance from its equilibrium position. The region in which the electron was originally found then contains a net positive charge. There is, therefore, an electrostatic force on the electron. If we allow the electron to move, it will be accelerated toward its original equilibrium position. When the electron arrives at its equilibrium position, it will have gained a finite energy from the induced electric field and therefore will overshoot the position of zero force. The motion of the electron will be similar to that of the pendulum in a grandfather clock. The electron, in the absence of damping forces, will oscillate with constant amplitude about its equilibrium position. The picture we have formed of the zero-temperature oscillations is qualitatively correct though greatly oversimplified. Any charge unbalance will cause a large number of electrons to move since the electric field induced by the displacement of one electron acts on all electrons. Thus a large number of electrons will oscillate *collectively* at the electron plasma frequency. These zero-temperature longitudinal electron oscillations do not propagate. If account is taken of the finite temperature of the plasma, the thermal motion of the electrons will cause the longitudinal electron oscillations to propagate. In addition, the frequency of the electron oscillations will be changed slightly from the zero-temperature value.

We shall discuss the properties of the longitudinal electron plasma oscillations both from a hydrodynamic and from a Boltzmann equation viewpoint. Using the hydrodynamic approach, we shall consider the zero-temperature case. Using the Boltzmann equation, we shall determine the change in the characteristics of the electron plasma oscillations due to the finite temperature of the electrons.

11-2 The Hydrodynamic Approach: Zero-temperature Theory

We shall make the following assumptions in the hydrodynamic derivation of longitudinal electron plasma oscillations: (1) The electrons are at zero temperature; therefore the electron pressure is zero. (2) There are no collisions; therefore $\nu_{ce} = 0$. (3) The ions are stationary. The large ion mass makes the ion velocity negligible compared with the electron velocity for a given force field. (4) All quantities vary as $\exp[i(\mathbf{k}\cdot\mathbf{r} - \omega t)]$. (5) $\mathbf{B} = 0$. (6) The electron velocity and electron density variation are small.

Using assumptions (1), (2), and (5), we can write the momentum transfer equation for the electrons [Eq. (4-10)] as

$$\frac{d\mathbf{v}_e}{dt} = -\frac{|e|}{m_e}\mathbf{E} \tag{11-1}$$

The total derivative of $\mathbf{v}_e$ with respect to t can be expanded as follows:

$$\frac{d\mathbf{v}_e}{dt} = \frac{\partial \mathbf{v}_e}{\partial t} + (\mathbf{v}_e \cdot \nabla)\mathbf{v}_e \tag{11-2}$$

Since we assume the electron velocity to be small, the second-order term in velocity can be neglected with respect to the first-order term. Equation (11-1) then becomes

$$\frac{\partial \mathbf{v}_e}{\partial t} = -\frac{|e|}{m_e}\mathbf{E} \tag{11-3}$$

Substitution of the assumed time variation of $\mathbf{v}_e$ into Eq. (11-3) yields

$$i\omega \mathbf{v}_e = \frac{|e|}{m_e}\mathbf{E} \tag{11-4}$$

The current density, neglecting ion motion and assuming that the electron density variation is small, can be written

$$\mathbf{j} = -|e|n_{e0}\mathbf{v}_e \tag{11-5}$$

where n_{e0} is the electron density in the absence of oscillations. The electron velocity can be eliminated between Eqs. (11-4) and (11-5).

$$\mathbf{j} = -\frac{|e|^2 n_{e0}}{i\omega m_e}\mathbf{E} \tag{11-6}$$

The divergence equation for the electric field [Eq. (4-1*a*)], with $\mathbf{D} = \epsilon_0\mathbf{E}$, is

$$\nabla \cdot \mathbf{E} = \frac{\rho}{\epsilon_0}$$

which, after substitution of the assumed spatial variation, becomes

$$i\mathbf{k} \cdot \mathbf{E} = \frac{\rho}{\epsilon_0} \tag{11-7}$$

If we take the divergence of both sides of Eq. (4-2*a*), recalling that $(\nabla \cdot \nabla \times \mathbf{H})$ is zero, we find

$$\nabla \cdot \mathbf{j} + \nabla \cdot \frac{\partial \mathbf{D}}{\partial t} = 0 \tag{11-8}$$

After an interchange of time and space derivatives in Eq. (11-8) and after the elimination of $\mathbf{D}$ by the use of Eq. (4-1*a*), we arrive at the continuity equation for charge

$$\nabla \cdot \mathbf{j} + \frac{\partial \rho}{\partial t} = 0 \tag{11-9}$$

With the assumed time and spatial variation, Eq. (11-9) becomes

$$\mathbf{k} \cdot \mathbf{j} - \omega\rho = 0 \tag{11-10}$$

a result which we have seen before.

If we combine Eqs. (11-6), (11-7), and (11-10) so that $\mathbf{j}$ and ρ are eliminated, we find that

$$\left(1 - \frac{\omega_{pe}^2}{\omega^2}\right)\mathbf{k} \cdot \mathbf{E} = 0 \tag{11-11}$$

where ω_{pe} is the electron plasma frequency defined in Chap. 7,

$$\omega_{pe}^2 \equiv \frac{|e|^2 n_{e0}}{m_e \epsilon_0}$$

In order that Eq. (11-11) be satisfied, it is necessary that either (1) $\mathbf{k} \cdot \mathbf{E} = 0$ or (2) $\omega = \omega_{pe}$. The latter case implies longitudinal electron oscillations, as we shall soon see. The former implies a type of propagation which we have already discussed. Let us first dispose of the former alternative.

If $\mathbf{k} \cdot \mathbf{E} = 0$, the electric field is perpendicular to the direction of propagation, and, from Eq. (11-7), there is no net charge density within the plasma. Substitution of Eq. (11-6) into Eq. (4-2a) yields

$$\nabla \times \mathbf{H} = -i\omega\epsilon_0\left(1 - \frac{\omega_{pe}^2}{\omega^2}\right)\mathbf{E} \tag{11-12}$$

Thus the effective permittivity of the plasma is

$$\epsilon_p = \epsilon_0\left(1 - \frac{\omega_{pe}^2}{\omega^2}\right) \tag{11-13}$$

the result found in Sec. 7-3-1.

If $\omega = \omega_{pe}$, $\mathbf{k}$ and $\mathbf{E}$ can have components in the same direction. This type of oscillation is longitudinal. Since we can regard Eq. (11-11) as a dispersion relation, ω and $\mathbf{k}$ are unrelated. The longitudinal electron plasma oscillations have zero group velocity; the waves do not propagate. The behavior of the longitudinal electron plasma oscillations is such that the electrons oscillate back and forth about their equilibrium positions at a frequency equal to the electron plasma frequency.

11-3 The Boltzmann Equation Approach: Finite-temperature Theory

We shall make the following assumptions in this more rigorous derivation of the properties of the longitudinal electron plasma oscillations: (1) There are no collisions. (2) The ions are stationary. (3) Only variation in the x direction need be considered; all variable quantities vary as $\exp[i(k_x x - \omega t)]$. (4) $\mathbf{B} = 0$. (5) The oscillations are of small amplitude so that a "one-dimensional" distribution function for the electrons can be written

$$f_e(x,u_x,t) = f_0(u_x) + f_1(x,u_x,t) \tag{11-14}$$

where f_0 is the equilibrium electron distribution function and f_1 is a small perturbation on f_0. The distribution function is "one-dimensional" in the sense that $\int_{u_x}\int_x f_e(x,u_x,t)\,du_x\,dx = N$, the total number of electrons in the system. We now use the word "equilibrium" to signify that f_0 is not a function of time or of position coordinates. Note that this usage of the word "equilibrium" is different from that of Chap. 2.

With these assumptions the Boltzmann equation for electrons [Eq. (4-20)] can be written

$$\frac{\partial f_0}{\partial t} + \frac{\partial f_1}{\partial t} + u_x\frac{\partial f_0}{\partial x} + u_x\frac{\partial f_1}{\partial x} - \frac{|e|E_x}{m_e}\frac{\partial f_0}{\partial u_x} - \frac{|e|E_x}{m_e}\frac{\partial f_1}{\partial u_x} = 0 \tag{11-15}$$

where E_x is the induced electric field. Since f_0 is not a function of time or position coordinates, the first and third terms on the left-hand side of Eq. (11-15) are zero. The last term on the left of Eq. (11-15) contains the product of two small quantities E_x and f_1 and thus can be neglected. The "linearized" Boltzmann equation is therefore

$$\frac{\partial f_1}{\partial t} + u_x\frac{\partial f_1}{\partial x} - \frac{|e|E_x}{m_e}\frac{\partial f_0}{\partial u_x} = 0 \tag{11-16}$$

After substitution of the assumed time and spatial variation of f_1 and E_x into Eq. (11-16), we obtain

$$-i\omega f_1 + ik_x u_x f_1 - \frac{|e|E_x}{m_e}\frac{\partial f_0}{\partial u_x} = 0 \tag{11-17}$$

Equation (11-17) can be solved for the perturbation distribution function f_1.

$$f_1 = \frac{1}{i(k_x u_x - \omega)} \frac{|e|E_x}{m_e} \frac{\partial f_0}{\partial u_x} \tag{11-18}$$

We shall now write down the electric field divergence with the charge density expressed in terms of the distribution function.

$$\frac{\partial E_x}{\partial x} = \frac{Z_i|e|n_{i0}}{\epsilon_0} - \frac{|e|}{\epsilon_0}\left(\int_{u_x} f_0\, du_x + \int_{u_x} f_1\, du_x\right) \tag{11-19}$$

where we assume that $\mathbf{D} = \epsilon_0 \mathbf{E}$. Since $n_{e0} = \int_{u_x} f_0\, du_x$, and since the plasma is essentially charge-neutral in equilibrium, Eq. (11-19) becomes

$$\frac{\partial E_x}{\partial x} = -\frac{|e|}{\epsilon_0}\int_{u_x} f_1\, du_x \tag{11-20}$$

or, with the assumed spatial variation,

$$ik_x E_x = -\frac{|e|}{\epsilon_0}\int_{u_x} f_1\, du_x \tag{11-21}$$

Substitution of Eq. (11-18) into Eq. (11-21) yields the dispersion relation for any given equilibrium distribution function $f_0(u_x)$.

$$\frac{k_x^2}{\omega_{pe}^2} = \int_{u_x} \frac{1}{n_{e0}}\left(\frac{\partial f_0}{\partial u_x}\right)\left(\frac{1}{u_x - (\omega/k_x)}\right) du_x \tag{11-22}$$

The integral on the right of Eq. (11-22) has a pole at $u_x = \omega/k_x$. The integral can be evaluated by complex variable techniques. We shall adopt a simpler approach. We shall assume that the equilibrium distribution function is such that $\partial f_0/\partial u_x = 0$ when $u_x = \omega/k_x$. Saying the same thing in another way, we shall assume that ω/k_x is large compared with the dominant velocities in f_0. With this assumption, the denominator of the integral can be expanded in an infinite series. Equation (11-22) then becomes

$$\frac{k_x^2}{\omega_{pe}^2} = -\frac{k_x}{\omega}\int_{u_x} \frac{1}{n_{e0}} \frac{\partial f_0}{\partial u_x}\left[1 + \frac{u_x k_x}{\omega} + \left(\frac{u_x k_x}{\omega}\right)^2 + \left(\frac{u_x k_x}{\omega}\right)^3\right] du_x \tag{11-23}$$

where we have retained only the first four terms in the expansion. Each of the four integrals on the right of Eq. (11-23) can be integrated by parts. As an example, we shall integrate the last term on the right.

$$-\int_{-\infty}^{+\infty} \frac{1}{n_{e0}} \left(\frac{k_x}{\omega}\right)^4 \frac{\partial f_0}{\partial u_x} u_x^3 \, du_x = -\frac{1}{n_{e0}} \left(\frac{k_x}{\omega}\right)^4 u_x^3 f_0 \bigg]_{-\infty}^{+\infty} + \frac{1}{n_{e0}} \left(\frac{k_x}{\omega}\right)^4 \int_{-\infty}^{+\infty} 3u_x^2 f_0 \, du_x$$

$$= 0 + 3\left(\frac{k_x}{\omega}\right)^4 \overline{u_x^2}$$

The quantity $u_x^3 f_0$ is zero at $u_x = \pm\infty$ because f_0 decreases more rapidly with increasing u_x than u_x^3 increases. The equilibrium distribution function will, in general, decrease exponentially as u_x goes to infinity. After a similar integration by parts of the remaining three terms in Eq. (11-23), we find that Eq. (11-23) becomes

$$\frac{k_x^2}{\omega_{pe}^2} = \frac{k_x^2}{\omega^2}\left(1 + \frac{3k_x^2}{\omega^2}\overline{u_x^2}\right) \tag{11-24}$$

where $\overline{u_x^2}$ is the average value of the square of the velocity in the x direction. Since the second term on the right of Eq. (11-24) is small compared with one, the degree of approximation is not changed by substituting ω_{pe}^2 for ω^2 in that term. Equation (11-24) can then be solved for the frequency of the electron oscillations.

$$\omega^2 = \omega_{pe}^2\left(1 + \frac{3k_x^2\overline{u_x^2}}{\omega_{pe}^2}\right) \tag{11-25}$$

The frequency of the longitudinal electron oscillations is approximately the electron plasma frequency for small values of k_x. As k_x increases, as the wavelength decreases, the frequency of the electron oscillations increases. At the same time, however, the expansion used in the derivation begins to break down. As was expected, the addition to the frequency term derived using the Boltzmann equation is caused by the thermal motion of the electrons, $\overline{u_x^2}$. It is also this additional term which relates the propagation constant to the frequency and thus predicts a propagating wave.

One very important comment is in order. We have integrated Eq. (11-22) in a special way. We have ignored the pole in the integrand by assuming that there is also a zero at the same velocity. Strictly speaking, this procedure is not valid. By ignoring the pole, we have ignored the possibility of growing and decaying waves in the plasma. By proper integration of Eq. (11-22), the existence of Landau damping and of unstable electron oscillations may be predicted. The growing oscillations provide a basis for the design of microwave plasma tubes. We shall have more to say about growing and decaying waves in Chap. 17.

PROBLEMS

11-1 Show that Eq. (11-11) follows from Eq. (7-19). Discuss.

11-2 Show that if both ions and electrons are free to move under the influence of an electric field and if neither ions nor electrons have any thermal energy, the dispersion relation becomes

$$\left(1 - \frac{\omega_{pe}^2}{\omega^2} - \frac{\omega_{pi}^2}{\omega^2}\right) \mathbf{k} \cdot \mathbf{E} = 0 \tag{11-26}$$

11-3 Find numerical values for the electron plasma frequency as a function of electron density. Consider electron densities from 10^{16} electrons per cubic meter to 10^{22} electrons per cubic meter. Compare the electron plasma frequency with the electron–neutral-particle collision frequency if $n_e = n_n$, $\overline{\mathcal{E}_n} = .03$ ev, and (*a*) $\overline{\mathcal{E}_e} = 1$ ev, $\sigma_{en} = 10^{-19}$ m^2; (*b*) $\overline{\mathcal{E}_e} = 100$ ev, $\sigma_{en} = 10^{-20}$ m^2. (See Prob. 10-3.)

11-4 Show that the expansion used to obtain Eq. (11-25) is not valid if the wavelength of the longitudinal electron oscillations is not much greater than the Debye length. Assume that f_0 is a Maxwell-Boltzmann distribution function. Discuss.

11-5 Using the hydrodynamic equations and the ideal gas law, show that

$$\frac{k_x^2}{\omega_{pe}^2} = \frac{k_x^2}{\omega^2}\left(1 + \frac{\kappa T}{m}\frac{k_x^2}{\omega_{pe}^2}\right) \tag{11-27}$$

In order to obtain Eq. (11-27) it is necessary to assume the following:

(*a*) The ions are stationary; n_i = constant.

(*b*) Electron velocity, electron density variation, and induced electric field are small; products of small quantities can be neglected.

(*c*) All variable quantities vary as exp $[i(k_x x - \omega t)]$.

(*d*) The equation of state is $P_e = n_e \kappa T$.

The following relations, therefore, describe the longitudinal plasma oscillations from the fluid viewpoint:

$$m\frac{\partial \mathbf{v}_e}{\partial t} = -|e|\mathbf{E} - \frac{\nabla P_e}{n_e} \tag{11-28}$$

$$\frac{\partial n_e}{\partial t} + \nabla \cdot (n_e \mathbf{v}_e) = 0 \tag{11-29}$$

$$\nabla \cdot \mathbf{E} = \frac{\rho}{\epsilon_0} \tag{11-30}$$

$$\rho = (Z_i|e|n_i - |e|n_e) \tag{11-31}$$

$$\mathbf{j} = -|e|n_e\mathbf{v}_e \tag{11-32}$$

$$P_e = n_e \kappa T \tag{11-33}$$

Proceed by taking the divergence of Eq. (11-28). By proper substitution and use of assumption (*b*), a linear equation in ρ may be obtained. The dispersion relation is then found by using assumption (*c*). Compare Eq. (11-27) with Eq. (11-24).

HINT: The equation of state for an *adiabatic* gas in one-dimensional compression is

$$P = 3n\kappa T$$

11-6 Calculate the phase velocity and the group velocity of the longitudinal electron waves.

BIBLIOGRAPHY

Dawson, J.: On Landau Damping, *Phys. Fluids*, **4**:869 (1961).
On a calculation of the dynamical details of the electron behavior during Landau damping.

Gabor, L.: Plasma Oscillations, *Brit. J. Appl. Phys.*, **2**:209 (1951).
A survey of experimental and theoretical research in electron oscillations.

Jackson, J. D.: Plasma Oscillations, *J. Nucl. Energy, pt. C, Plasma-Accelerators-Thermonucl. Res.*, **1**:171 (1960).
On the Boltzmann equation approach to longitudinal electron oscillations.

Landau, L.: On the Vibrations of the Electronic Plasma, *J. Phys. (USSR)*, **10**:25 (1946).
On the integration of Eq. (11-22) by complex variable techniques.

Oster, L.: Linearized Theory of Plasma Oscillations, *Rev. Mod. Phys.*, **32**:141 (1960).
On the hydrodynamic versus the Boltzmann equation approach to electron oscillations.

Tonks, L., and I. Langmuir: Oscillations in Ionized Gases, *Phys. Rev.*, **33**:195 (1929).
On the hydrodynamic approach to longitudinal electron oscillations.

Vlasov, A. A.: On the Kinetic Theory of an Assembly of Particles with Collective Interaction, *J. Phys. (USSR)*, **9**:25 (1945).
On the integration of Eq. (11-22) as done in the chapter.

chapter

12

Longitudinal Oscillations of Plasma Ions

12-1 Introduction

Since the electrons in a plasma exhibit longitudinal oscillations, it is reasonable to expect that the other charged particle constituent of the plasma, the ions, will exhibit a similar type of oscillation. Studies of low-pressure gas discharges in the late 1920s revealed that in addition to the high-frequency oscillations caused by the plasma electrons, low-frequency oscillations were present. These low-frequency oscillations occur in a frequency band from zero to a few megacycles per second. The longitudinal oscillations of plasma electrons occur at a higher frequency than the corresponding ion oscillations because of the greater mobility of the electrons. Since electrons have much smaller masses than ions, the acceleration of the electrons in a given force field is much greater than that of the ions. We have seen in Chap. 11 that the ions play a very small role in determining the properties of the longitudinal electron oscillations. The converse is not true. The electrons, because of their high mobility, will play a significant part in determining the properties of the ion oscillations. Thus we should expect that the ion oscillations will differ from the electron oscillations in some respects. The two types of longitudinal oscillations have in common the fact that the induced electric fields which support the oscillations are parallel to the direction of the particle average motion and to the direction of propagation.

12-2 Derivation

Several times in the course of our investigations we have discussed what happens in differential volumes. Hopefully, the reader has achieved some degree of familiarity with the technique of setting up mathematical relationships by considering what is going on physically within a small volume of the plasma. We shall now consider the last small volume in the book. It is shown in Fig. 12-1. We shall assume that each ion is displaced from its equilibrium position by a *small* distance $\xi(x,t)$. Our goal in this derivation will be to obtain an equation for $\xi(x,t)$.

The number of ions entering the left-hand side of the differential volume d^3r shown in Fig. 12-1 is $n_i\xi\, dy\, dz$. The number leaving the right-hand side of the volume is given by the usual Taylor expansion of the number entering.

$$n_i\xi\, dy\, dz + \frac{\partial}{\partial x}(n_i\xi)\, dx\, dy\, dz$$

Since ξ is assumed to be only a function of the x coordinate (not of the y coordinate or z coordinate), no ions enter or leave d^3r through the other four faces. The net increase in the ion density in d^3r is therefore

$$\delta n_i = -\frac{\partial}{\partial x}(n_i\xi) \tag{12-1}$$

If the ions are displaced from their equilibrium positions, the redistribution of charge will set up electric fields within the plasma. The electrons will move in these fields. We shall assume that the electrons can move so rapidly that they follow any potential field set up by the ion disturbance and that the electron densities obey Maxwell-Boltzmann statistics $n_e = n_{e0} \exp(|e|V/\kappa T_e)$, where n_{e0} is the equilibrium electron density, $V(x)$ is the electric potential within the

Fig. 12-1 A differential volume of plasma.

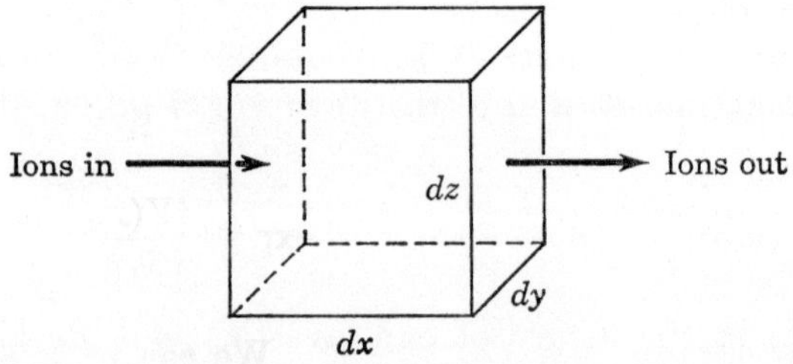

plasma, and T_e is the electron temperature. Thus the net increase in electron density in d^3r is given by

$$\delta n_e = n_{e0}\left[\exp\left(\frac{|e|V}{\kappa T_e}\right) - 1\right] \tag{12-2}$$

Since the plasma is electrically neutral in equilibrium, the net charge density within d^3r is the difference between the increase in number densities of the ions and of the electrons multiplied by their respective charges.

$$\rho = Z_i|e|\,\delta n_i - |e|\,\delta n_e \tag{12-3}$$

Equations (12-1) to (12-3) can be combined with Eq. (4-1*a*) to yield, in the one-dimensional case,

$$\frac{\partial E_x}{\partial x} = -\frac{|e|}{\epsilon_0}\left\{Z_i\frac{\partial}{\partial x}(n_i\xi) + n_{e0}\left[\exp\left(\frac{|e|V(x)}{\kappa T_e}\right) - 1\right]\right\} \tag{12-4}$$

where we assume that $\mathbf{D} = \epsilon_0\mathbf{E}$.

We can write the ion density as $n_i = (n_{i0} + n_{i1})$, where n_{i0} is the equilibrium value of ion density and n_{i1} represents the *small* change in ion density caused by the *small* displacement ξ. If $n_{i0} + n_{i1}$ is substituted into the first term on the right of Eq. (12-4), that term becomes

$$-\frac{|e|}{\epsilon_0}Z_i\frac{\partial}{\partial x}(n_i\xi) = -\frac{|e|}{\epsilon_0}Z_in_{i0}\frac{\partial\xi}{\partial x} - \frac{|e|}{\epsilon_0}Z_i\xi\frac{\partial n_{i1}}{\partial x} - \frac{|e|}{\epsilon_0}Z_in_{i1}\frac{\partial\xi}{\partial x} \tag{12-5}$$

The second and third terms on the right of Eq. (12-5) contain the products of the small quantities ξ and n_{i1}. These two terms can therefore be ignored in comparison with the first term on the right of Eq. (12-5). It follows that we can write

$$-\frac{|e|}{\epsilon_0}Z_i\frac{\partial}{\partial x}(n_i\xi) = -\frac{|e|}{\epsilon_0}Z_in_{i0}\frac{\partial\xi}{\partial x} \tag{12-6}$$

Equation (12-4) therefore becomes

$$\frac{\partial E_x}{\partial x} = -\frac{|e|}{\epsilon_0}\left\{Z_in_{i0}\frac{\partial\xi}{\partial x} + n_{e0}\left[\exp\left(\frac{|e|V(x)}{\kappa T_e}\right) - 1\right]\right\} \tag{12-7}$$

We shall now assume that $\kappa T_e > |e|V$. We can use this assumption and the condition of charge neutrality $Z_i n_{i0} = n_{e0}$ to simplify Eq. (12-7).

$$\frac{\partial E_x}{\partial x} = -\frac{Z_i|e|n_{i0}}{\epsilon_0}\left(\frac{\partial \xi}{\partial x} + \frac{|e|V(x)}{\kappa T_e}\right) \tag{12-8}$$

After taking the partial derivative of Eq. (12-8) with respect to x, recalling that $E_x = -\frac{\partial V}{\partial x}$, we find that

$$\frac{\partial^2 E_x}{\partial x^2} = -\frac{Z_i|e|n_{i0}}{\epsilon_0}\left(\frac{\partial^2 \xi}{\partial x^2} - \frac{|e|E_x}{\kappa T_e}\right) \tag{12-9}$$

The momentum transfer equation for the ions [Eq. (4-11)], assuming $\mathbf{B} = 0$, $\nu_{ci} = 0$, $\nabla P_i = 0$, and assuming the ion disturbance to be small, becomes

$$m_i \frac{\partial^2 \xi}{\partial t^2} = Z_i|e|E_x \tag{12-10}$$

Equation (12-10) can be differentiated twice with respect to x to yield

$$\frac{m_i}{Z_i|e|}\frac{\partial^2}{\partial x^2}\frac{\partial^2 \xi}{\partial t^2} = \frac{\partial^2 E_x}{\partial x^2} \tag{12-11}$$

Equations (12-9) to (12-11) can be combined to eliminate E_x. The resulting relation is

$$\frac{m_i}{Z_i|e|}\frac{\partial^2}{\partial x^2}\frac{\partial^2 \xi}{\partial t^2} = -\frac{Z_i|e|n_{i0}}{\epsilon_0}\left(\frac{\partial^2 \xi}{\partial x^2} - \frac{m_i}{Z_i\kappa T_e}\frac{\partial^2 \xi}{\partial t^2}\right) \tag{12-12}$$

or

$$\frac{\partial^2}{\partial x^2}\left(\frac{\partial^2 \xi}{\partial t^2} + \frac{Z_i^2|e|^2 n_{i0}}{m_i\epsilon_0}\xi\right) = \frac{Z_i|e|^2 n_{i0}}{\epsilon_0\kappa T_e}\frac{\partial^2 \xi}{\partial t^2} \tag{12-13}$$

Equation (12-13) is the equation which predicts the longitudinal ion oscillations.

Since we are looking for oscillations or traveling waves, it is reasonable to assume that ξ varies as

$$\xi = \xi_0 \exp\left[i(k_x x - \omega t)\right] \tag{12-14}$$

The substitution of Eq. (12-14) into the equation of motion [Eq. (12-13)] yields the dispersion relation for the ion waves, the relation of k_x to ω. The results of this substitution are

$$k_x^2\omega^2\xi_0 - \frac{Z_i^2|e|^2 n_{i0}}{m_i\epsilon_0}k_x^2\xi_0 = -\frac{Z_i|e|^2 n_{i0}\omega^2}{\epsilon_0\kappa T_e}\xi_0 \tag{12-15}$$

and therefore the propagation constant is given by

$$k_x = \left[\frac{Z_i|e|^2 n_{i0}/\epsilon_0 \kappa T_e}{(\omega_{pi}/\omega)^2 - 1}\right]^{1/2} \tag{12-16}$$

where ω_{pi}, the ion plasma frequency, is defined as

$$\omega_{pi}^2 \equiv \frac{Z_i^2|e|^2 n_{i0}}{m_i \epsilon_0} \tag{12-17}$$

in analogy with the electron plasma frequency.

The propagation constant is real for all frequencies less than the ion plasma frequency. Thus ion waves have an upper cutoff frequency. For propagation frequencies near the ion plasma frequency, the wavelength $2\pi/k_x$ of the oscillations will be very small. For very low-frequency oscillations, the propagation constant is small, and the wavelength is very long. The phase and group velocities of the ion waves can be calculated from Eqs. (7-5) and (7-6), respectively. For $\omega \ll \omega_{pi}$, $v_p \rightarrow (Z_i \kappa T_e/m_i)^{1/2}$. The wave exhibits the characteristics of a sound wave in a medium whose particles have the temperature of the electrons and the mass of the ions. The velocity of the wave is independent of the density of the particles. This type of wave is sometimes referred to as an "electrostatic sound wave."

Physically, we might expect the ion oscillations not to propagate; we might expect them to behave like the zero-temperature longitudinal electron plasma oscillations. We would be wrong. The rapid motion of the electrons rushing in to negate the fields set up by the ion unbalance causes the ion oscillations to propagate. In the case of the electron oscillations, the ions remain approximately stationary.

12-3 Remarks

In the derivation for the longitudinal ion oscillations, we have not used the hydrodynamic continuity equation for the ions [Eq. (4-14)] explicitly. We have, however, included the information contained in the continuity equation by considering the ion motion into and out of the differential volume d^3r. It is possible to obtain the behavior of the ion oscillations by two approaches different from the one we have used: (1) by solving the hydrodynamic equations (momentum transfer, continuity, and state equations) with the pressure terms included and (2) by solving the Boltzmann equation. The latter approach predicts the appearance of a damping mechanism which is not predicted by either hydrodynamic approach.

PROBLEMS

12-1 List all assumptions made in deriving the propagation constant for longitudinal ion oscillations. Discuss.

12-2 Find analytical expressions for the group and phase velocities of ion waves.

12-3 Show that for $\omega \ll \omega_{pi}$

$$v_p \rightarrow \left(\frac{Z_i \kappa T_e}{m_i}\right)^{1/2}$$

Under these conditions, find a numerical value for the phase velocity if the ions are singly ionized oxygen ions and the electron average energy is 10 ev.

12-4 Plot ω_{pi} as a function of ion density for densities between 10^{16} ions per cubic meter and 10^{22} ions per cubic meter. Assume that the ions are singly ionized oxygen.

12-5 Show that Eq. (12-1) is equivalent to the ion continuity equation.

BIBLIOGRAPHY

Bernstein, I. B., and R. M. Kulsrud: Ion Wave Instabilities, *Phys. Fluids,* **3**:937 (1960).
On the Boltzmann equation approach to longitudinal ion oscillations.

Crawford, F. W., and G. S. Kino: Oscillations and Noise in Low Pressure D-C Discharges, Microwave Laboratory Report no. 817, Stanford University, 1961; and *Proc. IRE,* **49**:1767 (1961).
Contains a derivation of the properties of longitudinal ion oscillations similar to that presented in the chapter and a general discussion, with references, of the merits of the various derivations of the properties of ion oscillations.

Fried, B. D., and R. W. Gould: Longitudinal Ion Oscillations in a Hot Plasma, *Phys. Fluids,* **4**:139 (1961).
On the Boltzmann equation approach to longitudinal ion oscillations.

Spitzer, L., Jr.: "Physics of Fully Ionized Gases," 2d ed., Interscience Publishers, Inc., New York, 1962.
sec. 3-2—Electrostatic Waves with No Magnetic Field. On the solution of the hydrodynamic equations, with pressure terms included, to obtain the properties of the longitudinal ion oscillations.

Tonks, L., and I. Langmuir: Oscillations in Ionized Gases, *Phys. Rev.,* **33**:195 (1929).
Contains a derivation of the properties of longitudinal ion oscillations similar to that presented in the chapter.

chapter

13

Hydromagnetic Waves in Plasmas

13-1 Introduction

If a collisionless plasma is immersed in a strong uniform magnetostatic field, the system of plasma and static magnetic field can support low-frequency electromagnetic waves whose velocity of propagation can be many orders of magnitude less than the speed of light in vacuum. These low-frequency low-velocity electromagnetic waves are known as "magnetohydrodynamic" or "hydromagnetic" waves. The frequency of these electromagnetic waves must be less than the cyclotron frequency of the plasma ions if the waves are to display "hydromagnetic" characteristics.

We shall discuss two types of hydromagnetic waves: (1) those whose time-varying magnetic field is parallel to the static magnetic field and (2) those whose time-varying magnetic field is perpendicular to the static magnetic field. For both we assume that the transverse time-varying magnetic fields of the hydromagnetic wave are small compared with the static uniform magnetic field B_{0z} in which the plasma is immersed. In other words, we shall consider the hydromagnetic waves to be small-amplitude plane waves.

We shall examine the properties of hydromagnetic wave propagation from the viewpoint of orbit theory. We shall first combine the appropriate particle current densities discussed in Chap. 6 with Maxwell's equations. Maxwell's equations will then be solved for the dispersion relation.

The validity of orbit theory to describe mathematically the occurrence of hydromagnetic waves should certainly be questioned. In Chap. 6 we discussed the necessary conditions for the validity of orbit theory. Briefly, the magnetic and electric fields must not vary in time or space over the orbit of any one particle. In the present situation, therefore, the time-varying magnetic fields must be very small or perturbational in comparison with the static magnetic field B_{0z}, and the frequency of the time-varying fields must be much smaller than the cyclotron frequency of any charged particle in the system. Under these conditions the magnetic dipole moments of the charged particles will remain constant, and the current densities due to the various drifts of the magnetic dipole moments will be as listed in Chap. 6.

13-2 Magnetoacoustic Waves

We consider now the case of a hydromagnetic wave propagating in the y direction with its time-varying magnetic field B_z in the same direction as the uniform static magnetic field B_{0z}. Since we are concerned with a plane wave propagating in the y direction, the time-varying fields E_x and B_z can be assumed to vary as $\exp [i(k_y y - \omega t)]$. A diagram of the system under consideration is shown in Fig. 13-1. For a plane wave, the time-varying magnetic field is a spatial function of only the direction of propagation. There will therefore be no magnetic mirror fields present and no motion of the guiding centers of the type indicated by Eq. (5-42). If we assume charge neutrality, the external field drift caused by the electric field of the wave will result in no net electrical current since ions and electrons will move in the same direction. The current contribution given by Eq. (6-3) is therefore equal to zero. The remaining current densities which we must consider are (1) the curvature-drift current density [Eq. (6-2)], (2) the gradient-drift current density [Eq. (6-1)], (3) the polarization-

Fig. 13-1 The fields of the magnetoacoustic wave.

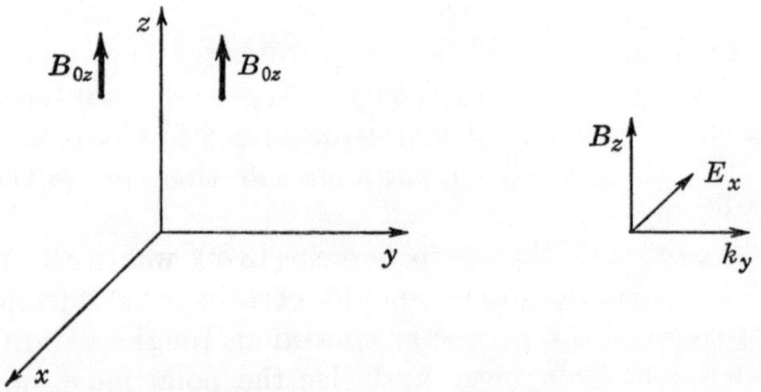

drift current density [Eq. (6-4)], and (4) the magnetization current density [Eq. (6-5)].

$$\mathbf{j}_c = 0 \tag{13-1}$$

$$\mathbf{j}_G = \left(\frac{n_i\overline{\mathcal{U}_i}}{|\mathbf{B}|} + \frac{n_e\overline{\mathcal{U}_e}}{|\mathbf{B}|}\right)\mathbf{a}_z \times \nabla_\perp|\mathbf{B}| \tag{13-2}$$

$$\mathbf{j}_P = \frac{n_im_i + n_em_e}{|\mathbf{B}|^2}\frac{\partial\mathbf{E}_\perp}{\partial t} \tag{13-3}$$

$$\mathbf{j}_M = -\nabla \times \frac{\overline{\mathcal{U}_i}n_i\mathbf{B}}{|\mathbf{B}|} - \nabla \times \frac{\overline{\mathcal{U}_e}n_e\mathbf{B}}{|\mathbf{B}|} \tag{13-4}$$

There are no curvature drifts because the magnetic field is always in the z direction.

If we make use of the assumed spatial variation, the gradient-drift current density [Eq. (13-2)] can be written

$$\mathbf{j}_G = -\frac{iW_\perp}{B_{0z}^2}k_yB_z\mathbf{a}_x \tag{13-5}$$

where $W_\perp \equiv (n_i\overline{\mathcal{E}_{i\perp}} + n_e\overline{\mathcal{E}_{e\perp}})$. That Eq. (13-5) follows from Eq. (13-2) can be shown as follows: The absolute value of the magnetic field is given by

$$|\mathbf{B}| = B_{0z} + B_z \tag{13-6}$$

since there is no B_y or B_x. It follows then that

$$\nabla_\perp|\mathbf{B}| = \frac{\partial B_z}{\partial x}\mathbf{a}_x + \frac{\partial B_z}{\partial y}\mathbf{a}_y \tag{13-7}$$

Since B_z is only a function of y and varies as e^{ik_yy}, Eq. (13-7) becomes

$$\nabla_\perp|\mathbf{B}| = ik_yB_z\mathbf{a}_y \tag{13-8}$$

After crossing a unit vector in the z direction into Eq. (13-8), we find that

$$\mathbf{a}_z \times \nabla_\perp|\mathbf{B}| = -ik_yB_z\mathbf{a}_x \tag{13-9}$$

If Eq. (13-9) is substituted into Eq. (13-2) with $\overline{\mathcal{U}} = \overline{\mathcal{E}_\perp}/|\mathbf{B}|$ and $|\mathbf{B}| \cong B_{0z}$, the resulting equation is Eq. (13-5).

Using the assumed time variation and making the approximations that $m_i \gg m_e$, $|\mathbf{B}| \cong B_{0z}$, we can write the polarization-drift current density [Eq. (13-3)] as

$$\mathbf{j}_P = -\frac{n_i m_i}{B_{0z}^2} i\omega \mathbf{E}_\perp \tag{13-10}$$

The magnetization current density [Eq. (13-4)] can be written

$$\mathbf{j}_M = -\frac{iW_\perp}{B_{0z}^2} k_y B_z \mathbf{a}_x \tag{13-11}$$

Equation (13-11) is valid if the following three conditions are satisfied: (1) $\bar{\mathfrak{u}}$ is not a function of position coordinates, (2)

$$\frac{n_i}{|\mathbf{B}|} = \text{constant} \qquad \frac{n_e}{|\mathbf{B}|} = \text{constant} \tag{13-12}$$

and (3) $|\mathbf{B}| \cong B_{0z}$. Since the magnetic moment of each individual plasma particle remains constant, it is reasonable to assume that the average magnetic moment is independent of position. We have proved in Chap. 6 [Eq. (6-37)] that Eq. (13-12) is true for the fields under consideration.

Equations (4-2*a*) and (4-3*a*) are the equations from which the properties of electromagnetic waves are derived.

$$\nabla \times \mathbf{H} = \mathbf{j} + \frac{\partial \mathbf{D}}{\partial t} \tag{4-2a}$$

$$\nabla \times \mathbf{E} = -\frac{\partial \mathbf{B}}{\partial t} \tag{4-3a}$$

The current density appearing in Eq. (4-2*a*) is the sum of the current densities given by Eqs. (13-5), (13-10), and (13-11). Equation (4-2*a*) can therefore be written

$$\frac{k_y B_z}{\mu_0} = -\frac{W_\perp k_y B_z}{B_{0z}^2} - \frac{W_\perp k_y B_z}{B_{0z}^2} - \frac{\omega n_i m_i}{B_{0z}^2} E_x - \omega\epsilon_0 E_x \tag{13-13}$$

where $\mathbf{D} = \epsilon_0 \mathbf{E}$ and $\mathbf{B} = \mu_0 \mathbf{H}$. Equation (13-13) can be rewritten

$$k_y \left(\frac{1}{\mu_0} + \frac{2W_\perp}{B_{0z}^2} \right) B_z + \left(\frac{\omega m_i n_i}{B_{0z}^2} + \omega\epsilon_0 \right) E_x = 0 \tag{13-14}$$

Equation (4-3*a*), for the assumed fields and field variation, becomes

$$\omega B_z + k_y E_x = 0 \tag{13-15}$$

Equations (13-14) and (13-15) can be solved simultaneously for k_y as a function of ω by setting the determinant of the matrix multiplying E_x, B_z equal to zero.

$$\left\| \begin{matrix} k_y \left(\dfrac{1}{\mu_0} + \dfrac{2W_\perp}{B_{0z}^2} \right) & \left(\dfrac{\omega m_i n_i}{B_{0z}^2} + \omega \epsilon_0 \right) \\ \omega & k_y \end{matrix} \right\| = 0 \tag{13-16}$$

The solution of Eq. (13-16) is

$$k_y^2 = \frac{\omega^2}{c^2} \frac{1 + n_i m_i/\epsilon_0 B_{0z}^2}{1 + W_\perp/(B_{0z}^2/2\mu_0)} \tag{13-17}$$

which is the dispersion relation with $c^2 = 1/\mu_0\epsilon_0$. The term $W_\perp$ has the same dimensions as the magnetic pressure $B_{0z}^2/2\mu_0$. $W_\perp$ can therefore be considered to be the kinetic pressure perpendicular to the z direction (see Chap. 8, Sec. 8-3).

The phase velocity and the group velocity of the hydromagnetic wave can be found by employing Eqs. (7-5) and (7-6), respectively.

$$v_g = v_p = c \left[\frac{1 + W_\perp/(B_{0z}^2/2\mu_0)}{1 + n_i m_i/\epsilon_0 B_{0z}^2} \right]^{1/2} \tag{13-18}$$

It is instructive to note that as the ion and the electron densities become small (n_i, $W_\perp \to 0$), the phase and group velocities of the hydromagnetic wave approach the speed of light. The wave becomes a low-frequency electromagnetic wave in free space. The case of interest to us is that for which $n_i m_i/\epsilon_0 B_{0z}^2 \gg 1$. Under these circumstances, Eq. (13-18) becomes

$$v_g = v_p = \left[\frac{2(W_\perp + B_{0z}^2/2\mu_0)}{n_i m_i} \right]^{1/2} \tag{13-19}$$

If we consider $(W_\perp + B_{0z}^2/2\mu_0)$ to be the effective total pressure of the system and $n_i m_i$ to be the effective mass density of the system, Eq. (13-19) represents the form of the velocity of an acoustic wave. Further, as we shall see in the next paragraph, the variation of the plasma particle density in the presence of the hydromagnetic wave under consideration is very similar to the density variation which occurs in a sound wave. For these reasons a hydromagnetic wave whose time-varying magnetic field is parallel to the uniform static magnetic field is often called a magnetoacoustic wave.

If the magnetoacoustic wave propagates through a system of plasma and uniform static magnetic field, both the total magnetic field and the ion and electron densities suffer compressions and rarefactions as a function of the coordinate associated with the direction of propagation and of time. We shall now show that this is the case. Let us write the perturbational magnetic field as

$$B_z = B_{1z}e^{i(k_y y - \omega t)}$$

The magnitude of the total magnetic field is then

$$|\mathbf{B}| = B_{0z} + B_{1z}e^{i(k_y y - \omega t)}$$

Equation (13-12), which follows from Eq. (6-37), can be written

$$\begin{aligned} n_i(y,t) &= C_i(B_{0z} + B_{1z}e^{i(k_y y - \omega t)}) \\ n_e(y,t) &= C_e(B_{0z} + B_{1z}e^{i(k_y y - \omega t)}) \end{aligned}$$

where the C's are constants. The particle densities are composed of a constant component plus a component which is directly proportional to the perturbational magnetic field B_z. The ion and electron densities behave in exactly the same manner as the total magnetic field. When the total magnetic field suffers a rarefaction, the particle densities suffer a proportional rarefaction. When the total magnetic field suffers a compression, the particle densities suffer a proportional compression.

We can examine the variation in the particle densities which occurs when a magnetoacoustic wave propagates through a system of plasma and uniform static magnetic field from a physical point of view. The three current densities which we have considered in the derivation of the magnetoacoustic wave are all in the x direction and hence cannot lead to a particle density variation. It is therefore the mass flow caused by the crossed time-varying electric and static magnetic fields which yields the density variation. Since this mass drift does not produce an electrical current, it was not included in Eq. (13-13). The mass flow due to the effects of the time-varying electric field and the static magnetic field is in the y direction. The density variation has the same frequency and spatial characteristics as the electric field which causes it. We see from Eq. (13-15) that the electric and magnetic fields of the magnetoacoustic wave are related to each other by a constant factor. Thus the density variation must also have the same characteristics as the magnetic field variation. The particles behave as if they were attached to the field lines of the total magnetic field.

When the field lines become more dense—when the total field becomes stronger—the particle densities also become greater. When the field lines become less dense—when the total field becomes weaker—the particle densities also decrease.

In obtaining the properties of the magnetoacoustic wave, we have used the fact that

$$\frac{d}{dt}\frac{n}{|\mathbf{B}|} = 0 \tag{6-37}$$

We have seen in Chap. 6 that Eq. (6-37) is a consequence of the continuity property of the ions and of the electrons. Thus, in obtaining the behavior of the magnetoacoustic wave, we have, in effect, made use of the continuity equations, Maxwell's equations, and the guiding-center drift velocities as determined from orbit theory.

13-3 Alfvén Waves

We consider now the case in which propagation is in the z direction and the varying field quantities are E_x, E_y, B_x, B_y. There will be no drifts due to magnetic mirror fields because B_x and B_y are not functions of x and y, and $B_z = 0$. There will be no electrical current densities due to gradient field drifts because the magnetic field in the z direction B_{0z} is a constant. There will, however, be an electrical current density due to curvature drifts since B_x and B_y are functions of z. In addition there will be a magnetization current density and a polarization current density.

The polarization current density [Eq. (13-3)] can be approximated as

$$\mathbf{j}_P = \frac{n_i m_i}{B_{0z}^2}\frac{\partial \mathbf{E}_\perp}{\partial t} \tag{13-20}$$

The magnetization current density is given by Eq. (13-4), in which we assume $\mathcal{U}$ to be independent of position, the particle densities to be constant, and $|\mathbf{B}| \cong B_{0z}$.

$$\mathbf{j}_M = -\frac{W_\perp}{B_{0z}^2}\nabla \times \mathbf{B} \tag{13-21}$$

The particle densities are constant because each guiding-center drift velocity is perpendicular to the direction of propagation. The current density due to curvature drifts is

$$\mathbf{j}_c = \frac{2(\overline{\mathcal{E}_{i\parallel}}n_i + \overline{\mathcal{E}_{e\parallel}}n_e)}{|\mathbf{B}|}\mathbf{a}_B \times [(\mathbf{a}_B \cdot \nabla)\mathbf{a}_B] \tag{6-2}$$

We shall now show that Eq. (6-2) can, in the present case, be written

$$\mathbf{j}_c = \frac{2W_{\|}}{B_{0z}^2} \nabla \times \mathbf{B} \tag{13-22}$$

where $W_{\|} \equiv (\overline{\mathcal{E}_{i\|}} n_i + \overline{\mathcal{E}_{e\|}} n_e)$. We begin by writing down the following vector identity:

$$(\mathbf{a}_B \cdot \nabla)\mathbf{a}_B = -\mathbf{a}_B \times (\nabla \times \mathbf{a}_B) + \tfrac{1}{2}\nabla|\mathbf{a}_B|^2 \tag{13-23}$$

The second term on the right of Eq. (13-23) is zero since $|\mathbf{a}_B|^2 = 1$.

$$(\mathbf{a}_B \cdot \nabla)\mathbf{a}_B = -\mathbf{a}_B \times (\nabla \times \mathbf{a}_B) \tag{13-24}$$

We can cross $\mathbf{a}_B$ into both sides of Eq. (13-24) and simplify the right-hand side.

$$\mathbf{a}_B \times [(\mathbf{a}_B \cdot \nabla)\mathbf{a}_B] = -\mathbf{a}_B \times [\mathbf{a}_B \times (\nabla \times \mathbf{a}_B)] = +\nabla \times \mathbf{a}_B \tag{13-25}$$

Equation (6-2) can therefore be written

$$\mathbf{j}_c = \frac{2W_{\|}}{|\mathbf{B}|} \nabla \times \frac{\mathbf{B}}{|\mathbf{B}|} \tag{13-26}$$

Since the perturbational magnetic fields of the hydromagnetic wave are perpendicular to the static field in the z direction, the absolute value of $\mathbf{B}$ is essentially the constant B_{0z}. Alternatively, $|\mathbf{B}|^{-1}\nabla \times (\mathbf{B}/|\mathbf{B}|)$ can be expanded, with the second-order terms in B_x and B_y being ignored, to yield the same result. If $|\mathbf{B}|$ is set equal to B_{0z} in Eq. (13-26), the resulting equation is Eq. (13-22).

We can combine Eq. (4-2*a*) with the appropriate current densities [Eqs. (13-20) to (13-22)] to yield

$$\nabla \times \mathbf{H} = \frac{2W_{\|}\mu_0}{B_{0z}^2} \nabla \times \mathbf{H} - \frac{W_{\perp}\mu_0}{B_{0z}^2} \nabla \times \mathbf{H} + \frac{n_i m_i}{B_{0z}^2}\frac{\partial \mathbf{E}_{\perp}}{\partial t} + \epsilon_0 \frac{\partial \mathbf{E}_{\perp}}{\partial t} \tag{13-27}$$

where again $\mathbf{B} = \mu_0\mathbf{H}$, $\mathbf{D} = \epsilon_0\mathbf{E}$. Equation (13-27) can be cast into the following form:

$$\nabla \times \mathbf{H} = \frac{\epsilon_0(1 + n_i m_i/\epsilon_0 B_{0z}^2)}{1 + \mu_0 W_{\perp}/B_{0z}^2 - 2W_{\|}\mu_0/B_{0z}^2} \frac{\partial \mathbf{E}_{\perp}}{\partial t} \tag{13-28}$$

The factor multiplying $\partial\mathbf{E}_{\perp}/\partial t$ is the effective permittivity of the system of plasma and static magnetic field. The effective dielectric constant of the system is therefore

$$K_p = \frac{1 + n_i m_i/\epsilon_0 B_{0z}^2}{1 + \mu_0 W_\perp/B_{0z}^2 - 2W_\parallel \mu_0/B_{0z}^2} \tag{13-29}$$

We shall consider the case for which $n_i m_i/\epsilon_0 B_{0z}^2 \gg 1$. Under these circumstances, Eq. (13-29) reduces to

$$K_p = \frac{c^2 n_i m_i}{W_\perp - 2W_\parallel + B_{0z}^2/\mu_0} \tag{13-30}$$

The relationship between the effective dielectric constant and the propagation constant of an isotropic medium is given by Eq. (7-23).

$$k = \frac{\omega}{c}\sqrt{K} \tag{7-23}$$

It follows that

$$k_z = \omega \left(\frac{n_i m_i}{W_\perp - 2W_\parallel + B_{0z}^2/\mu_0} \right)^{1/2} \tag{13-31}$$

The phase velocity and the group velocity of the hydromagnetic wave are

$$v_g = v_p = \left(\frac{W_\perp - 2W_\parallel + B_{0z}^2/\mu_0}{n_i m_i} \right)^{1/2} \tag{13-32}$$

The type of hydromagnetic wave which we are now considering, the type which propagates in the direction parallel to B_{0z}, was first treated by Alfvén. This type of hydromagnetic wave is therefore called an "Alfvén wave." In an Alfvén wave all guiding-center drifts are perpendicular to the direction of propagation. Thus the Alfvén wave can induce no variation in the plasma particle densities. If $W_\parallel < [(W_\perp/2) + (B_{0z}^2/2\mu_0)]$, the behavior of the Alfvén wave is similar to the behavior of the magnetoacoustic wave with the exception that the Alfvén wave does not modulate the density of the plasma particles.

If $W_\parallel > [(W_\perp/2) + (B_{0z}^2/2\mu_0)]$, the dielectric constant [Eq. (13-30)] is negative. If we assume k real, Eq. (13-31) yields an angular frequency ω which is a positive or negative imaginary number. Thus the amplitude of the Alfvén wave may increase exponentially with time. The physical reason for this instability is the following: Particles moving along the curved magnetic field lines exert a centrifugal force which tends to distort the field lines further. There are, however, two effects which tend to straighten out the field lines. These effects are caused by the magnetic pressure and by the magnetization current density. If the centrifugal force dominates, the wave amplitude grows exponentially; if not, the wave amplitude oscillates. If the Alfvén wave grows, the fields will reach a magnitude at which our various approximations become invalid. It is then impossible for us to predict the behavior of the wave by using the techniques of orbit theory.

It is sometimes convenient to think of $2W_{\parallel}$ as the kinetic pressure of the particles in the z direction, just as $W_{\perp}$ can be thought of as the kinetic pressure component perpendicular to the z direction. Thus it is the kinetic pressure in the z direction which causes the instability. The instability is inhibited by the magnetic pressure and by the kinetic pressure perpendicular to the z direction.

13-4 Alternative Derivations

We have obtained the properties of magnetoacoustic waves and of Alfvén waves by using the techniques of orbit theory. Many of the results which we have derived in this chapter can be obtained by alternative techniques.

In Chap. 7 we considered the interaction of electromagnetic waves with plasmas. Since hydromagnetic waves are low-frequency electromagnetic waves, the mathematical approach used in Chap. 7 can be used to describe the properties of the hydromagnetic waves. In particular, a dispersion relation for Alfvén waves can be found by setting $\omega \ll \omega_{bi}$ in Eq. (7-34), and a dispersion relation for magnetoacoustic waves can be found by setting $\omega \ll \omega_{bi}$ in the dispersion relation of Prob. 7-7.

Further, the properties of the hydromagnetic waves can be determined without resort to the laborious mathematics of Chap. 7. It is possible to combine the momentum transfer equations with Maxwell's equations without actually solving the momentum transfer equations if it is assumed that the plasma is a perfect conductor. For this approach the reader is referred to the bibliography.

PROBLEMS

13-1 Under what conditions is $n_i m_i/\epsilon_0 B_{0z}^2 > 1$? Consider magnetic fields from 10^{-5} webers/m^2 to 1 weber/m^2.

13-2 Find an approximate numerical value for the propagation velocity of a magnetoacoustic wave traveling in the ionosphere.

13-3 Find a dispersion relation for Alfvén waves by letting $\omega \ll \omega_{bi}$ in Eq. (7-34). Does the derivation of Alfvén waves using the techniques of Chap. 7 place any restrictions on the magnitude of the plane waves?

13-4 Find a dispersion relation for magnetoacoustic waves by letting $\omega \ll \omega_{bi}$ in the results of Prob. 7-7. Does the derivation of magnetoacoustic waves using the techniques of Chap. 7 place any restrictions on the magnitude of the plane waves?

13-5 List all approximations made in deriving the properties of hydromagnetic waves using the techniques of orbit theory. List all approximations made in deriving the properties of hydromagnetic waves using the techniques of Chap. 7. Which technique yields the more detailed view of the hydromagnetic waves? Discuss the differences in the dispersion relations found from orbit theory and those found in Probs. 13-3 and 13-4.

13-6 Using the techniques of orbit theory, determine the dispersion relation for the propagation of a hydromagnetic wave in the y direction if the following fields are present: B_{0z}, B_x, B_z, E_x, E_z. Note that there will be both curvature and gradient drifts.

13-7 Using the techniques of orbit theory, formulate Maxwell's equations for Alfvén waves and for magnetoacoustic waves without making the approximation $|\mathbf{B}| \cong B_{0z}$. As the final step before obtaining the dispersion relations, expand the $|\mathbf{B}|$'s which appear in the equations in power series. In the resultant equations retain only the first-order terms. Compare these results with Eqs. (13-14) and (13-27).

BIBLIOGRAPHY

Alfvén, H.: "Cosmical Electrodynamics," Clarendon Press, Oxford, 1950.
chap. iv—On the properties and derivation of hydromagnetic waves.

Astrom, E.: On Waves in an Ionized Gas, *Arkiv Fysik*, **2**:443 (1950).
On the derivation of hydromagnetic waves from the viewpoint of Chap. 7.

Chandrasekhar, S.: "Plasma Physics," The University of Chicago Press, Chicago, 1960.
p. 146—On hydromagnetic waves from the viewpoint of orbit theory.

Glasstone, S., and R. H. Lovberg: "Controlled Thermonuclear Reactions," D. Van Nostrand Company, Inc., Princeton, N.J., 1960.
sec. 4.114—Hydromagnetic Waves. On the mathematical approach to hydromagnetic waves discussed in the last paragraph of Sec. 13-4.

Longmire, C. L.: "Elementary Plasma Physics," Interscience Publishers, Inc., New York, 1963.
chap. 6—Small-amplitude Plasma Waves. Contains a discussion of hydromagnetic waves from the viewpoint of orbit theory.

Rose, D. J., and M. Clark, Jr.: "Plasmas and Controlled Fusion," John Wiley & Sons, Inc., New York, 1961.
sec. 9.4—On the derivation of Alfvén waves from the viewpoint of Chap. 7.

Spitzer, L., Jr.: "Physics of Fully Ionized Gases," 2d ed., Interscience Publishers, Inc., New York, 1962.
sec. 3.3—Hydromagnetic Waves. On the mathematical approach to hydromagnetic waves discussed in the last paragraph of Sec. 13.4, plus a general discussion of hydromagnetic waves.

chapter

14

Diffusion, Mobility, and Transport Phenomena

14-1 Introduction

If we could follow the motion of one charged particle within a plasma, we would find the exact trajectory of that particle to be very complex. Let us consider first the case for which there are no external forces acting on the particle. If the collision interactions in the plasma are primarily short-range (electron-neutral or ion-neutral collisions), the particle trajectory will be a series of straight segments during which the particle moves at its thermal velocity joined by very small regions of collision interaction where the direction of the thermal velocity is altered. The trajectory of a single particle which suffers only short-range collisions is shown in Fig. 14-1. If the collision interactions in the plasma are primarily long-range (Coulomb collisions), the individual plasma particle will, in general, move randomly but will not move in a straight path since it will always be undergoing a collision

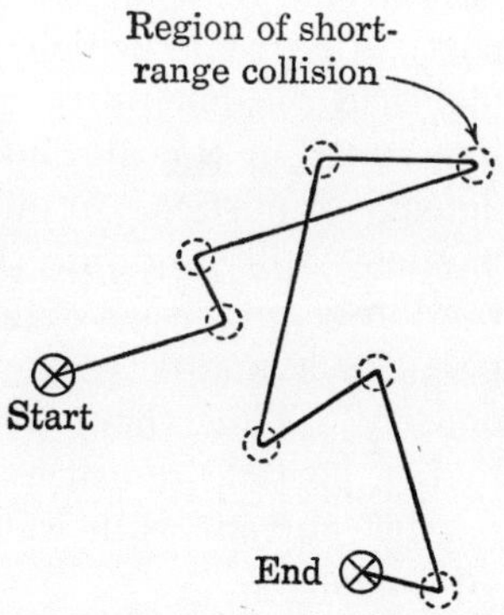

Fig. 14-1 The trajectory of a particle which suffers only short-range collisions.

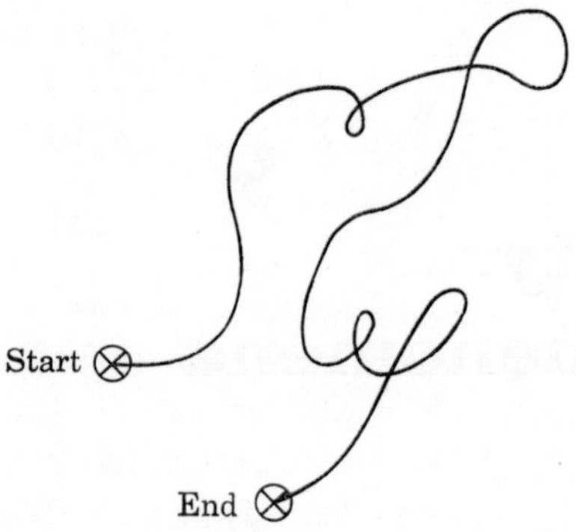

Fig. 14-2 The trajectory of a particle which suffers only long-range collisions.

interaction. The trajectory of a single particle which suffers only long-range collisions is shown in Fig. 14-2. Whether the collision interactions are short-range or long-range or a combination of the two, the individual particle will move randomly about in the plasma because of a combination of its thermal motion and its collision interactions.

If a force is applied to the charged particle whose trajectory we are examining, that trajectory will be altered from its behavior in the absence of a force. As the particle attempts to move with its thermal velocity, its trajectory is warped in the direction of the applied force. The trajectory of a charged particle under the influence of an applied force is shown in Fig. 14-3, where we assume the collision interactions to be short-range. The net effect of the applied force is to cause the particle to move in the direction of the force. Note that although the force in Fig. 14-3 is shown constant in direction, there is a force (the magnetic force) which does not act in a constant direction.

Now that we have examined qualitatively the motion of an individual charged plasma particle, we are in a position to discuss the particle currents which may arise from the motion of a system of identical plasma particles. Let us consider first the currents which may arise in the absence of an applied force. If there is a greater density of particles in one region of the plasma than in another, more particles will randomly leave the region of higher density than will randomly enter that region. This is so simply because there are more particles available to perform random motion in the region of higher density. Thus there will be a particle current out of the region of higher particle density. The process by which there can be a net particle current while each individual particle moves randomly is called "diffusion."

If an external force is applied to the system of identical plasma particles, the particles will exhibit a net motion in the direction of the applied force. The applied force can therefore cause a particle current density whose origin is different from that of the diffusion current. If the particles exhibit a net motion under the influence of the applied force, we can assign to the particles a "mobility." We shall mathematically define the mobility in Secs. 14-2 and 14-5.

In the preceding discussion we considered qualitatively the processes of diffusion and of force-induced motion. It is not, however, clear that the mathematics of the plasma motion will provide us with this distinction. In point of fact, the distinction between diffusion current and force-induced current can be shown mathematically only if the force-induced velocity is very small compared with the average thermal velocity of the particles.

The particle current which may exist because of the motion of the system of plasma particles which we have been considering represents a transport of mass and charge within the plasma. Mass transport and charge transport are the most elementary of the transport phenomena. As an individual plasma particle moves from place to place within the plasma, that particle will carry with it various other properties. One important property which may be transported by the plasma particles is kinetic energy (heat). For example, if the particles in one region of a plasma have higher thermal energies than

Fig. 14-3 The trajectory of a particle which suffers only short-range collisions in the presence of an applied force.

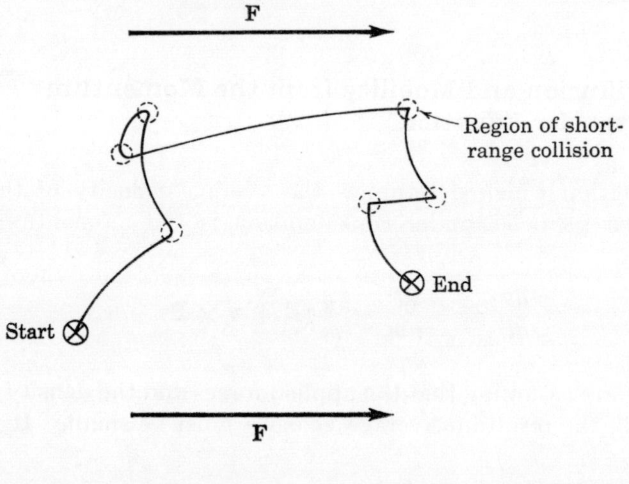

the particles in the remainder of the plasma, thermal energy will be transported out of the hot region by the moving particles. Similarly, any property of the plasma particles will be transported along with the particles as they move because of their own random velocities and the effects of applied forces.

In this chapter we shall study the transport properties of a system of identical charged plasma particles which satisfy the following conditions:

1. No property of the system is time-varying.
2. The system is only slightly disturbed from its equilibrium configuration: density gradients and applied forces are small.
3. The collision frequencies of the particles under study are independent of position coordinates but may be a function of particle speed.

There are two systems of plasma particles which satisfy the conditions given above. They are the ions and the electrons of a slightly ionized plasma in which the neutral-particle density is approximately constant. In a slightly ionized plasma the neutral-particle density greatly exceeds the charged-particle densities. Thus the charged particles collide primarily with neutral particles. The collision frequency of the charged particles cannot be a function of position if the neutral-particle density is uniform.

The results which we shall derive in this chapter will not be applicable to highly ionized plasmas because of the restrictions which we have placed on the collision frequencies.

The mathematical techniques used in this chapter are very similar to those used in the treatment of electron transport phenomena in metals and semiconductors.

14-2 Diffusion and Mobility from the Momentum Transfer Equation

The equation which determines the average velocity of the system of plasma particles under consideration is the momentum transfer equation

$$\frac{d\mathbf{v}}{dt} = -\frac{\nabla P}{nm} + \frac{q}{m}(\mathbf{E} + \mathbf{v} \times \mathbf{B}) - \nu_c \mathbf{v} \tag{14-1}$$

Since we are assuming that the applied forces and the density gradients are small, the resultant average velocity must be small. It is reasonable then to make the small-velocity approximation

$$\frac{d\mathbf{v}}{dt} = \frac{\partial\mathbf{v}}{\partial t} + (\mathbf{v}\cdot\nabla)\mathbf{v} \cong \frac{\partial\mathbf{v}}{\partial t} \tag{14-2}$$

We shall consider only an applied electric field. The momentum transfer equation for the time-invariant case therefore becomes

$$0 = -\frac{\nabla P}{nm} + \frac{q\mathbf{E}}{m} - \nu_c\mathbf{v} \tag{14-3}$$

Equation (14-3) can be rewritten

$$\mathbf{v} = -\frac{\nabla P}{nm\nu_c} + \frac{q\mathbf{E}}{m\nu_c} \tag{14-4}$$

The particle current density (not the electrical current density) is given by

$$\mathbf{\Gamma} = n\mathbf{v} = -\frac{\nabla P}{m\nu_c} + \frac{nq\mathbf{E}}{m\nu_c} \tag{14-5}$$

Since we are assuming that the equilibrium distribution function is not disturbed much by the applied electric field or by the small pressure gradient, the kinetic pressure of the system of particles under consideration can be written

$$P = \frac{nm}{3}\overline{u^2} \tag{3-13}$$

Substitution of Eq. (3-13) into Eq. (14-5) yields

$$\mathbf{\Gamma} = -\nabla\left(\frac{\overline{u^2}n}{3\nu_c}\right) + \frac{nq\mathbf{E}}{m\nu_c} \tag{14-6}$$

where we have brought the effective collision frequency ν_c inside the gradient sign. The effective collision frequency is assumed *not* to be a function of position coordinates.

The first term on the right of Eq. (14-6) is the diffusion current density

$$\mathbf{\Gamma}_{\text{diffusion}} = -\nabla(Dn)$$

where the diffusion coefficient D is defined as

$$D \equiv \overline{u^2}/3\nu_c \tag{14-7}$$

The diffusion coefficient of the system of particles is directly proportional to the average energy of the particles and inversely proportional to the uniform density of passive scattering centers (neutral particles). The diffusion current represents a motion of particles due to random processes from regions of greater particle concentration to regions of the lesser particle concentration. Even if the particle density is uniform, the variation in particle energy with position will yield a diffusion current. The reason for this is apparent from a physical point of view: Hot particles travel faster than cold particles, and thus hot particles will diffuse out of a region more quickly than cold particles of an equal density will diffuse in.

The second term on the right of Eq. (14-6) is the current density due to the effects of the applied electric field.

$$\mathbf{\Gamma}_{\text{mobility}} = n\mathfrak{u}\mathbf{E} \tag{14-8}$$

where the mobility is defined by Eqs. (14-8) and (14-6)

$$\mathfrak{u} \equiv \frac{q}{m\nu_c} \tag{14-9}$$

We have previously derived Eq. (14-8) in Sec. 5-3-2, although we did not use the word "mobility" in that section.

The total particle current density can be rewritten

$$\mathbf{\Gamma} = -\nabla(Dn) + n\mathfrak{u}\mathbf{E} \tag{14-10}$$

where D and $\mathfrak{u}$ are defined by Eqs. (14-7) and (14-9), respectively.

The ratio of the diffusion coefficient to the mobility is found by dividing Eq. (14-9) by Eq. (14-7).

$$\frac{D}{\mathfrak{u}} = \frac{m\overline{u^2}}{3q} \tag{14-11}$$

The ratio is independent of the effective collision frequency. Since $\overline{\mathcal{E}} = m\overline{u^2}/2$, the average energy, diffusion coefficient, and mobility of the system of particles under consideration are related by

$$\frac{D}{\mathfrak{u}} = \frac{2\overline{\mathcal{E}}}{3q} \tag{14-12}$$

If the equilibrium distribution function of the system of particles is a Maxwell-Boltzmann distribution, the average energy of the particles is essentially

$$\bar{\mathcal{E}} = \tfrac{3}{2}\kappa T \tag{2-16}$$

The substitution of Eq. (2-16) into Eq. (14-12) yields

$$\frac{D}{\mathrm{u}} = \frac{\kappa T}{q} \tag{14-13}$$

Equation (14-13) is known as the "Einstein relation."

14-3 Transport Theory: The Anisotropic Distribution Function

All properties of the system of plasma particles under study can be found from a solution of the Boltzmann equation

$$\frac{\partial f}{\partial t} + \mathbf{u}\cdot\nabla_r f + \frac{\mathbf{F}}{m}\cdot\nabla_u f = \left(\frac{\partial f}{\partial t}\right)_{\text{collisions}} \tag{14-14}$$

In order to be able to solve the Boltzmann equation, we must express the collision term in an analytical form. We shall assume that the right-hand side of Eq. (14-14) can be written

$$\left(\frac{\partial f}{\partial t}\right)_{\text{collisions}} = -(f - f_0)\nu_R \tag{14-15}$$

where $f_0(\mathbf{r},u)$ is an isotropic distribution function and where $\nu_R(u)$ is defined as the *relaxation collision frequency* of the system of particles.

We can advance an argument to show that Eq. (14-15) is reasonable. Let us assume that at $t = 0$ a *spatially uniform* system of particles under the influence of a force $\mathbf{F}$ has a distribution function $f(\mathbf{u})$. At $t = 0^+$, the force $\mathbf{F}$ is removed. The distribution function must therefore decay toward its isotropic value $f_0(u)$. For $t > 0^+$, Eqs. (14-14) and (14-15) can be combined to yield

$$\frac{\partial f}{\partial t} = -(f - f_0)\nu_R \tag{14-16}$$

The solution to Eq. (14-16) is

$$(f - f_0)_t = (f - f_0)_{t=0}e^{-\nu_R t} \tag{14-17}$$

The distribution function relaxes to its isotropic value with a time constant $\tau_R = 1/\nu_R$. From a physical point of view, the relaxation or decay of the distribution function toward its isotropic value must be caused by collision interactions. Hence it is reasonable to call ν_R the relaxation collision frequency. We shall show later that under certain conditions the relaxation collision frequency is identical with the effective collision frequency defined in connection with the momentum transfer equation. Since the use of the assumed collision term [Eq. (14-15)] yields physically reasonable results in the particular case we have examined, we can use the assumed collision term in other situations with some degree of confidence.

In the time-invariant case, the Boltzmann equation with the assumed collision term becomes

$$\mathbf{u} \cdot \nabla_r f + \frac{\mathbf{F}}{m} \cdot \nabla_u f = -(f - f_0)\nu_R \tag{14-18}$$

which we can alternatively write as

$$f = f_0 - \frac{1}{\nu_R}\left(\mathbf{u} \cdot \nabla_r f + \frac{\mathbf{F}}{m} \cdot \nabla_u f\right) \tag{14-19}$$

The second term on the right-hand side of Eq. (14-19) is small since it represents the deviation from the isotropic distribution function. Saying the same thing in another way, the second term on the right of Eq. (14-19) is small because the density gradient and the applied force are assumed small. Equation (14-19) can be solved by the method of successive approximations. The first approximation f^1 to the distribution function is

$$f^1 = f_0 \tag{14-20}$$

To obtain the second approximation to f, we insert f^1 into the right-hand side of Eq. (14-19).

$$f^2 = f_0 - \frac{1}{\nu_R}\left(\mathbf{u} \cdot \nabla_r f_0 + \frac{\mathbf{F}}{m} \cdot \nabla_u f_0\right) \tag{14-21}$$

To obtain the third approximation to f, we would substitute f^2 from Eq. (14-21) into the right-hand side of Eq. (14-19). We shall, however, use the second approximation to the distribution function.

$$f = f_0 - \frac{1}{\nu_R}\left(\mathbf{u} \cdot \nabla_r f_0 + \frac{\mathbf{F}}{m} \cdot \nabla_u f_0\right) \tag{14-22}$$

We shall now determine the mass and charge transport characteristics of the system of plasma particles under consideration. The particle current density at any point within the plasma is

$$\mathbf{\Gamma}(\mathbf{r}) = n\mathbf{v} = \iiint_{\text{all } \mathbf{u}} f\mathbf{u}\, d^3u \tag{14-23}$$

The substitution of Eq. (14-22) into Eq. (14-23) yields

$$\mathbf{\Gamma} = -\iiint_{\text{all } \mathbf{u}} \left(\frac{\mathbf{u}}{\nu_R} \cdot \nabla_r f_0 + \frac{\mathbf{F}}{m\nu_R} \cdot \nabla_u f_0 \right) \mathbf{u}\, d^3u \tag{14-24}$$

since

$$\iiint_{\text{all } \mathbf{u}} f_0 \mathbf{u}\, d^3u = 0 \tag{14-25}$$

The integral of the product of an even function of $\mathbf{u}$, $f_0(\mathbf{r},u)$, and an odd function of $\mathbf{u}$, $\mathbf{u}$, over all $\mathbf{u}$ is zero.

Equation (14-24) is an equation for the flow or transport of particles. The transport of mass is obtained by multiplying both sides of Eq. (14-24) by the mass m of the plasma particles under consideration. The transport of charge (the electrical current) is obtained by multiplying both sides of Eq. (14-24) by the charge q of the plasma particles under consideration.

The plasma particles move from place to place because of their own random velocities and the effects of the applied forces. Each of the plasma particles has certain properties which it will carry with it as it moves from place to place within the plasma. For example, each plasma particle has a kinetic energy. Let us calculate the transport or flow of kinetic energy connected with the motion of the plasma particles under consideration. The current density of the particles whose velocities are between $\mathbf{u}$ and $\mathbf{u} + d\mathbf{u}$ is

$$d\mathbf{\Gamma}_u = \mathbf{u} f\, d^3u \tag{14-26}$$

These particles have a kinetic energy $mu^2/2$. The energy or heat flow due to these particles is therefore

$$d\mathbf{Q}_u = (d\mathbf{\Gamma}_u)(mu^2/2) \tag{14-27}$$

The total heat flow is

$$\mathbf{Q}(\mathbf{r}) = \iiint_{\text{all } \mathbf{u}} f\mathbf{u}\tfrac{1}{2}mu^2\, d^3u \tag{14-28}$$

If Eq. (14-22) is substituted into Eq. (14-28), the resultant equation is

$$\mathbf{Q} = -\frac{m}{2}\iiint\limits_{\text{all }\mathbf{u}} \left(\frac{\mathbf{u}}{\nu_R}\cdot\nabla_r f_0 + \frac{\mathbf{F}}{\nu_R m}\cdot\nabla_u f_0\right)\mathbf{u}u^2\, d^3u \qquad (14\text{-}29)$$

where the integral over $f_0\mathbf{u}$ again yields zero.

In a similar way any transport property of the particles can be determined. If $\mathfrak{F}(\mathbf{u})$ is any scalar property of the particles, the flow of $\mathfrak{F}$ is

$$\mathbf{S}(\mathbf{r}) = \iiint\limits_{\text{all }\mathbf{u}} f\mathfrak{F}(\mathbf{u})\mathbf{u}\, d^3u \qquad (14\text{-}30)$$

If $\mathfrak{F}(\mathbf{u})$ is an even function of $\mathbf{u}$, the flow of $\mathfrak{F}$ is

$$\mathbf{S} = -\iiint\limits_{\text{all }\mathbf{u}} \left(\frac{\mathbf{u}}{\nu_R}\cdot\nabla_r f_0 + \frac{\mathbf{F}}{m\nu_R}\cdot\nabla_u f_0\right)\mathfrak{F}(\mathbf{u})\mathbf{u}\, d^3u \qquad (14\text{-}31)$$

since f_0 is an even function of $\mathbf{u}$. If $\mathfrak{F}(\mathbf{u})$ is an odd function of $\mathbf{u}$, the flow of $\mathfrak{F}$ is

$$\mathbf{S} = \iiint\limits_{\text{all }\mathbf{u}} f_0\mathfrak{F}(\mathbf{u})\mathbf{u}\, d^3u \qquad (14\text{-}32)$$

since
$$\left(\frac{\mathbf{u}}{\nu_R}\cdot\nabla_r f_0 + \frac{\mathbf{F}}{m\nu_R}\cdot\nabla_u f_0\right)$$

is an odd function of $\mathbf{u}$.

14-4 Diffusion from the Boltzmann Equation

In this section we shall examine the first integral on the right of Eq. (14-24), the diffusion current.

$$\mathbf{\Gamma}_{\text{diffusion}} = -\iiint\limits_{\text{all }\mathbf{u}} \left(\frac{\mathbf{u}}{\nu_R}\cdot\nabla_r f_0\right)\mathbf{u}\, d^3u \qquad (14\text{-}33)$$

The integrand of Eq. (14-33) can be expanded as

$$\left(\frac{u_x}{\nu_R}\frac{\partial f_0}{\partial x} + \frac{u_y}{\nu_R}\frac{\partial f_0}{\partial y} + \frac{u_z}{\nu_R}\frac{\partial f_0}{\partial z}\right)(u_x\mathbf{a}_x + u_y\mathbf{a}_y + u_z\mathbf{a}_z) \qquad (14\text{-}34)$$

When Eq. (14-34) is further expanded, six of the nine terms will contain odd powers of u_x, u_y, or u_z. The odd powers of u_x, u_y, or u_z are odd functions of $\mathbf{u}$. The terms $(1/\nu_R)(\partial f_0/\partial x)$, $(1/\nu_R)(\partial f_0/\partial y)$, and $(1/\nu_R)(\partial f_0/\partial z)$ are even functions of $\mathbf{u}$. It follows that the integral over the terms containing odd powers of u_x, u_y, or u_z will be zero. Equation (14-34) therefore reduces to

$$\mathbf{\Gamma}_{\text{diff}} = -\iiint\limits_{\text{all } \mathbf{u}} \left(\frac{u_x^2}{\nu_R} \frac{\partial f_0}{\partial x} \mathbf{a}_x + \frac{u_y^2}{\nu_R} \frac{\partial f_0}{\partial y} \mathbf{a}_y + \frac{u_z^2}{\nu_R} \frac{\partial f_0}{\partial z} \mathbf{a}_z \right) d^3u \tag{14-35}$$

We shall now show that

$$\iiint\limits_{\text{all } \mathbf{u}} \frac{u_x^2}{\nu_R} \frac{\partial f_0}{\partial x}\, d^3u = \iiint\limits_{\text{all } \mathbf{u}} \frac{u^2}{3\nu_R} \frac{\partial f_0}{\partial x}\, d^3u \tag{14-36}$$

$$\iiint\limits_{\text{all } \mathbf{u}} \frac{u_y^2}{\nu_R} \frac{\partial f_0}{\partial y}\, d^3u = \iiint\limits_{\text{all } \mathbf{u}} \frac{u^2}{3\nu_R} \frac{\partial f_0}{\partial y}\, d^3u \tag{14-37}$$

$$\iiint\limits_{\text{all } \mathbf{u}} \frac{u_z^2}{\nu_R} \frac{\partial f_0}{\partial z}\, d^3u = \iiint\limits_{\text{all } \mathbf{u}} \frac{u^2}{3\nu_R} \frac{\partial f_0}{\partial z}\, d^3u \tag{14-38}$$

Equations (14-36) to (14-38), once proved, will then be used to simplify Eq. (14-35). It is sufficient to prove one of the above three equations. We shall concentrate our attention on Eq. (14-36). Because $\partial f_0/\partial x$ and ν_R are functions of speed, not of velocity, it follows that

$$\iiint\limits_{\text{all } \mathbf{u}} \frac{u_x^2}{\nu_R} \frac{\partial f_0}{\partial x}\, d^3u = \iiint\limits_{\text{all } \mathbf{u}} \frac{u_y^2}{\nu_R} \frac{\partial f_0}{\partial x}\, d^3u = \iiint\limits_{\text{all } \mathbf{u}} \frac{u_z^2}{\nu_R} \frac{\partial f_0}{\partial x}\, d^3u \tag{14-39}$$

Since

$$u^2 = u_x^2 + u_y^2 + u_z^2 \tag{14-40}$$

the three equal terms appearing in Eq. (14-39) can be summed to yield

$$\begin{aligned} &\iiint\limits_{\text{all } \mathbf{u}} \frac{u_x^2}{\nu_R} \frac{\partial f_0}{\partial x}\, d^3u + \iiint\limits_{\text{all } \mathbf{u}} \frac{u_y^2}{\nu_R} \frac{\partial f_0}{\partial x}\, d^3u + \iiint\limits_{\text{all } \mathbf{u}} \frac{u_z^2}{\nu_R} \frac{\partial f_0}{\partial x}\, d^3u \\ &\qquad = \iiint\limits_{\text{all } \mathbf{u}} \frac{u^2}{\nu_R} \frac{\partial f_0}{\partial x}\, d^3u = 3 \iiint\limits_{\text{all } \mathbf{u}} \frac{u_x^2}{\nu_R} \frac{\partial f_0}{\partial x}\, d^3u \end{aligned} \tag{14-41}$$

We see from Eq. (14-41) that

$$\iiint\limits_{\text{all } \mathbf{u}} \frac{u_x^2}{\nu_R} \frac{\partial f_0}{\partial x}\, d^3u = \iiint\limits_{\text{all } \mathbf{u}} \frac{u^2}{3\nu_R} \frac{\partial f_0}{\partial x}\, d^3u \tag{14-36}$$

We have proved Eq. (14-36). The proof of Eqs. (14-37) and (14-38) follows in a similar way. The substitution of Eqs. (14-36) to (14-38) into Eq. (14-35) yields

$$\mathbf{\Gamma}_{\text{diff}} = -\iiint\limits_{\text{all } \mathbf{u}} \frac{u^2}{3\nu_R} \nabla_r f_0 \, d^3u \tag{14-42}$$

Since **r** and **u** are independent, the integration and differentiation operations can be interchanged with the result that

$$\mathbf{\Gamma}_{\text{diff}} = -\nabla \iiint\limits_{\text{all } \mathbf{u}} \frac{u^2 f_0}{3\nu_R} \, d^3u \equiv -\nabla(Dn) \tag{14-43}$$

where

$$D \equiv \overline{\left(\frac{u^2}{3\nu_R}\right)} \tag{14-44}$$

is the diffusion coefficient. Notice that we have assumed that ν_R is a function of u but not of **r**.

The diffusion coefficient of Eq. (14-44) may be compared with the diffusion coefficient of Eq. (14-7), which was derived using the momentum transfer equation. It is apparent that if ν_R is not a function of u, then the relaxation collision frequency is equal to the effective collision frequency.

14-5 Mobility from the Boltzmann Equation

In this section we shall examine the second integral on the right of Eq. (14-24), the particle current caused by the applied forces which act on the particles.

$$\mathbf{\Gamma}_{\text{mobility}} = -\iiint\limits_{\text{all } \mathbf{u}} \left(\frac{q\mathbf{E}}{m\nu_R} \cdot \nabla_u f_0\right) \mathbf{u} \, d^3u \tag{14-45}$$

where we have assumed that only an electric force $\mathbf{F} = q\mathbf{E}$ is present. Since

$$u^2 = u_x^2 + u_y^2 + u_z^2 \tag{14-40}$$

it follows that

$$\frac{\partial f_0}{\partial u_x} = \frac{u_x}{u}\frac{\partial f_0}{\partial u} \qquad \frac{\partial f_0}{\partial u_y} = \frac{u_y}{u}\frac{\partial f_0}{\partial u} \qquad \frac{\partial f_0}{\partial u_z} = \frac{u_z}{u}\frac{\partial f_0}{\partial u} \tag{14-46}$$

The integrand of Eq. (14-45) can therefore be written

$$\left(\frac{q}{m\nu_R}\right)\left(\frac{E_x u_x}{u}\frac{\partial f_0}{\partial u} + \frac{E_y u_y}{u}\frac{\partial f_0}{\partial u} + \frac{E_z u_z}{u}\frac{\partial f_0}{\partial u}\right)(u_x\mathbf{a}_x + u_y\mathbf{a}_y + u_z\mathbf{a}_z) \tag{14-47}$$

If Eq. (14-47) is further expanded, the terms in u_xu_y, u_xu_z, and u_yu_z will integrate to zero since $\partial f_0/\partial u$ is an even function of $\mathbf{u}$. Equation (14-45) becomes

$$\mathbf{\Gamma}_{\text{mobility}} = -\iiint_{\text{all }\mathbf{u}} \left(\frac{q}{muv_R}\right)(E_xu_x^2\mathbf{a}_x + E_yu_y^2\mathbf{a}_y + E_zu_z^2\mathbf{a}_z)\left(\frac{\partial f_0}{\partial u}\right) d^3u \tag{14-48}$$

We shall now show that

$$\iiint_{\text{all }\mathbf{u}} \frac{E_xu_x^2}{v_Ru}\frac{\partial f_0}{\partial u}\, d^3u = \iiint_{\text{all }\mathbf{u}} \frac{E_xu}{3v_R}\frac{\partial f_0}{\partial u}\, d^3u \tag{14-49}$$

$$\iiint_{\text{all }\mathbf{u}} \frac{E_yu_y^2}{v_Ru}\frac{\partial f_0}{\partial u}\, d^3u = \iiint_{\text{all }\mathbf{u}} \frac{E_yu}{3v_R}\frac{\partial f_0}{\partial u}\, d^3u \tag{14-50}$$

$$\iiint_{\text{all }\mathbf{u}} \frac{E_zu_z^2}{uv_R}\frac{\partial f_0}{\partial u}\, d^3u = \iiint_{\text{all }\mathbf{u}} \frac{E_zu}{3v_R}\frac{\partial f_0}{\partial u}\, d^3u \tag{14-51}$$

Equations (14-49) to (14-51), once proved, will then be used to simplify Eq. (14-48). Since $[(1/v_Ru)(\partial f_0/\partial u)]$ is a function of speed, not of velocity, it follows that

$$\iiint_{\text{all }\mathbf{u}} \frac{u_x^2}{uv_R}\frac{\partial f_0}{\partial u}\, d^3u = \iiint_{\text{all }\mathbf{u}} \frac{u_y^2}{uv_R}\frac{\partial f_0}{\partial u}\, d^3u = \iiint_{\text{all }\mathbf{u}} \frac{u_z^2}{uv_R}\frac{\partial f_0}{\partial u}\, d^3u \tag{14-52}$$

If we make use of Eq. (14-40), the three equal terms in Eq. (14-52) can be summed to yield

$$\iiint_{\text{all }\mathbf{u}} \frac{u_x^2}{uv_R}\frac{\partial f_0}{\partial u}\, d^3u + \iiint_{\text{all }\mathbf{u}} \frac{u_y^2}{uv_R}\frac{\partial f_0}{\partial u}\, d^3u + \iiint_{\text{all }\mathbf{u}} \frac{u_z^2}{uv_R}\frac{\partial f_0}{\partial u}\, d^3u = \iiint_{\text{all }\mathbf{u}} \frac{u}{v_R}\frac{\partial f_0}{\partial u}\, d^3u \tag{14-53}$$

Equations (14-49) to (14-51) follow from Eqs. (14-52) and (14-53). If Eqs. (14-49) to (14-51) are substituted into Eq. (14-48), Eq. (14-48) becomes

$$\mathbf{\Gamma}_{\text{mobility}} = -\mathbf{E}\iiint_{\text{all }\mathbf{u}} \frac{qu}{3mv_R}\frac{\partial f_0}{\partial u}\, d^3u \tag{14-54}$$

The integrand of Eq. (14-54) is only a function of u. The volume element d^3u can therefore be integrated over angular velocity coordinates with the result that

$$\mathbf{\Gamma}_{\text{mobility}} = -\mathbf{E}\int_0^\infty \frac{qu}{3mv_R}\frac{\partial f_0}{\partial u}\, 4\pi u^2\, du \tag{14-55}$$

Equation (14-55) can be integrated by parts.

$$\mathbf{\Gamma}_{\text{mobility}} = -\mathbf{E}\left(\frac{4\pi u^3}{3m\nu_R} f_0\right)\Big]_0^\infty + \mathbf{E}\int_0^\infty f_0 \frac{d}{du}\frac{4\pi q u^3}{3m\nu_R}\, du \tag{14-56}$$

The first term on the right of Eq. (14-56) is equal to zero since f_0 goes to zero faster than u^3 goes to infinity as u approaches infinity. The isotropic distribution function will, in general, decrease exponentially as u goes to infinity. Equation (14-56) therefore becomes

$$\mathbf{\Gamma}_{\text{mobility}} = \mathbf{E}\int_0^\infty f_0 \frac{d}{du}\frac{4\pi q u^3}{3m\nu_R}\, du \tag{14-57}$$

The mobility has been defined by the relationship

$$\mathbf{\Gamma}_{\text{mobility}} = n\mathfrak{u}\mathbf{E} \tag{14-8}$$

It follows that

$$\mathfrak{u} = \frac{1}{n}\int_0^\infty f_0 \frac{d}{du}\frac{4\pi q u^3}{3m\nu_R}\, du \tag{14-58}$$

If we consider the case for which ν_R is not a function of u, Eq. (14-58) can be expanded to yield

$$\mathfrak{u} = \frac{q}{nm\nu_R}\int_0^\infty f_0 4\pi u^2\, du \tag{14-59}$$

The density of the system of particles under consideration is

$$n(\mathbf{r}) = \int_0^\infty f_0 4\pi u^2\, du \tag{14-60}$$

since

$$\frac{1}{\nu_R}\left(\mathbf{u}\cdot\nabla_r f_0 + \frac{\mathbf{F}}{m}\cdot\nabla_u f_0\right)$$

is an odd function of $\mathbf{u}$. It follows that Eq. (14-59) becomes

$$\mathfrak{u} = \frac{q}{m\nu_R} \tag{14-61}$$

From a comparison of Eqs. (14-61) and (14-9), it is again apparent that if the relaxation collision frequency is a constant, then it is identical with the effective collision frequency.

14-6 Heat Flow from the Boltzmann Equation

The flow of kinetic energy or heat due to the motion of the plasma particles under study is given by Eq. (14-29).

$$\mathbf{Q} = -\frac{m}{2}\iiint_{\text{all }\mathbf{u}} \left(\frac{\mathbf{u}}{\nu_R}\cdot\nabla_r f_0 + \frac{\mathbf{F}}{m\nu_R}\cdot\nabla_u f_0\right)\mathbf{u}u^2\,d^3u \qquad (14\text{-}29)$$

We shall first consider heat flow in the x direction in the presence of an electric field. Later we shall generalize the results of our analysis to include three-dimensional heat flow. The x component of the heat flow is

$$Q_x = -\frac{m}{2}\iiint_{\text{all }\mathbf{u}} (u_x)(u_x^2+u_y^2+u_z^2)\left(\frac{u_x}{\nu_R}\frac{\partial f_0}{\partial x}+\frac{u_y}{\nu_R}\frac{\partial f_0}{\partial y}+\frac{u_z}{\nu_R}\frac{\partial f_0}{\partial z}\right.$$
$$\left.+\frac{qE_x}{m\nu_R}\frac{u_x}{u}\frac{\partial f_0}{\partial u}+\frac{qE_y}{m\nu_R}\frac{u_y}{u}\frac{\partial f_0}{\partial u}+\frac{qE_z}{m\nu_R}\frac{u_z}{u}\frac{\partial f_0}{\partial u}\right)d^3u \qquad (14\text{-}62)$$

If we multiply out the integrand of Eq. (14-62), all terms which contain odd powers of u_x, u_y, or u_z will integrate to zero since $\partial f_0/\partial x$, $\partial f_0/\partial y$, $\partial f_0/\partial z$, $(1/u\nu_R)(\partial f_0/\partial u)$ are even functions of $\mathbf{u}$. Equation (14-62) becomes

$$Q_x = -\frac{m}{2}\iiint_{\text{all }\mathbf{u}}\left(\frac{u_x^2u^2}{\nu_R}\frac{\partial f_0}{\partial x}+\frac{u_x^2u^2}{u}\frac{qE_x}{m\nu_R}\frac{\partial f_0}{\partial u}\right)d^3u \qquad (14\text{-}63)$$

We can set $u_x^2 = u^2/3$ in the integrand of Eq. (14-63). The validity of this procedure has been discussed in the preceding two sections. Equation (14-63), after an interchange of differentiation and integration operations, becomes

$$Q_x = -\frac{\partial}{\partial x}\iiint_{\text{all }\mathbf{u}}\frac{mu^4}{6\nu_R}f_0\,d^3u - E_x\iiint_{\text{all }\mathbf{u}}\frac{qu^3}{6\nu_R}\frac{\partial f_0}{\partial u}\,d^3u \qquad (14\text{-}64)$$

The second term on the right of Eq. (14-64) can be expressed in spherical velocity coordinates and integrated by parts with the result that

$$Q_x = -\frac{\partial}{\partial x}\iiint_{\text{all }\mathbf{u}}\frac{mu^4}{6\nu_R}f_0\,d^3u + E_x\int_0^\infty f_0\frac{d}{du}\frac{4\pi qu^5}{6\nu_R}\,du \qquad (14\text{-}65)$$

Equation (14-65) can be generalized to provide a vector expression for three-dimensional heat flow.

$$\mathbf{Q} = -\nabla \iiint\limits_{\text{all }\mathbf{u}} \frac{mu^4}{6\nu_R} f_0 \, d^3u + \mathbf{E} \int_0^\infty f_0 \frac{d}{du} \frac{4\pi q u^5}{6\nu_R} \, du \qquad (14\text{-}66)$$

The electric field which appears in Eq. (14-66) may be in part induced by the nonuniform density and nonuniform energy of the charged plasma particles and in part applied by external sources. We can eliminate the electric field from Eq. (14-66) by combining Eq. (14-66) with the equation for the total particle current density

$$\mathbf{\Gamma} = -\nabla(Dn) + n\mathrm{u}\mathbf{E} \qquad (14\text{-}67)$$

where the mobility is defined by Eq. (14-58) and the diffusion coefficient by Eq. (14-44). The resultant equation is

$$\mathbf{Q} = -\nabla \iiint\limits_{\text{all }\mathbf{u}} \frac{mu^4}{6\nu_R} f_0 \, d^3u + \frac{\mathbf{\Gamma} + \nabla(Dn)}{n\mathrm{u}} \int_0^\infty f_0 \frac{d}{du} \frac{4\pi q u^5}{6\nu_R} \, du \qquad (14\text{-}68)$$

Thus the heat flow can be calculated as a function of the particle current density. Very often it is useful to know the heat flow in a situation in which there is no closed path for current flow. The heat flow in these situations is found by setting $\mathbf{\Gamma} = 0$ in Eq. (14-68).

Equation (14-68) can be further simplified if the relaxation collision frequency is a constant. Under these conditions, Eq. (14-68) can be written

$$\mathbf{Q} = -\nabla \iiint\limits_{\text{all }\mathbf{u}} \frac{mu^4}{6\nu_R} f_0 \, d^3u + \frac{\mathbf{\Gamma} + \nabla(Dn)}{n\mathrm{u}\nu_R} \iiint\limits_{\text{all }\mathbf{u}} \frac{5}{6} q u^2 f_0 \, d^3u \qquad (14\text{-}69)$$

Equation (14-69) can be alternatively written

$$\mathbf{Q} = -\nabla \left(\frac{mn}{6\nu_R} \overline{u^4} \right) + \frac{\mathbf{\Gamma} + \nabla(Dn)}{\mathrm{u}\nu_R} \frac{5}{6} q \overline{u^2} \qquad (14\text{-}70)$$

The heat flow equation cannot be obtained from the momentum transfer equation. It can, however, be obtained from an *energy* transfer equation. The energy transfer equation is derived from the Boltzmann equation by multiplying each term in the Boltzmann equation by $mu^2/2$ and integrating over all possible velocities, just as the momentum transfer equation can be derived by multiplying each term in the Boltzmann equation by $m\mathbf{u}$ and integrating over all possible velocities (see Appendix A).

14-7 An Equation for Particle Density

We have shown that under the conditions set forth in Sec. 14-1, the particle current density of a system of identical particles can be written

$$\mathbf{\Gamma} = -\nabla(Dn) + n\mathbf{u}\mathbf{E} \tag{14-67}$$

The system of identical particles must also satisfy a continuity equation

$$\frac{\partial n}{\partial t} + \nabla \cdot \mathbf{\Gamma} = 0 \tag{14-71}$$

We can combine Eqs. (14-67) and (14-71) so as to eliminate the particle current density.

$$\frac{\partial n}{\partial t} - \nabla^2(Dn) + \nabla \cdot (n\mathbf{u}\mathbf{E}) = 0 \tag{14-72}$$

If D, $\mathbf{u}$, and $\mathbf{E}$ are known and if initial conditions and boundary conditions are given, Eq. (14-72) can be solved for the particle density as a function of time and of spatial coordinates.

In the absence of an electric field, Eq. (14-72) becomes

$$\nabla^2(Dn) = \frac{\partial n}{\partial t} \tag{14-73}$$

Equation (14-73) is called the "diffusion equation." If the diffusion coefficient is independent of position coordinates, Eq. (14-73) can be written

$$D\nabla^2 n = \frac{\partial n}{\partial t} \tag{14-74}$$

This form of the diffusion equation is particularly useful in the theory of electrons and holes in semiconductors.

14-8 Discussion

We stated in Sec. 14-1 that the results of this chapter would not be applicable to highly ionized plasmas. The reason for this is the following: In this chapter we have assumed that the system of particles under study (electrons or ions) collides primarily with other particles (neutrals) of *uniform* density. This assumption enabled us to employ effective and relaxation collision frequencies which were not dependent on position coordinates. In a highly ionized plasma the electrons and ions interact primarily with each other and with themselves. In an analysis of a highly ionized plasma employing the momentum transfer equations, the electron effective collision frequency must therefore take account of the behavior of the ions, and the ion effective collision frequency must take account of the behavior of the electrons. The momentum transfer equation for the ions is coupled to the momentum transfer equation for the electrons, and the two equations must be solved simultaneously. Further, when there is appreciable interaction between the charged plasma particles, the collision term in the Boltzmann equation for the electrons must take account of the behavior of the ions, and the collision term in the Boltzmann equation for the ions must take account of the behavior of the electrons. The two Boltzmann equations must be solved simultaneously. Thus, in general, to study transport phenomena in a highly ionized plasma it is necessary to solve two equations simultaneously.

We have defined the diffusion coefficient and the mobility for the case that the only applied force is that of an electric field. The extension of the mathematical approach we have employed to cover the case of both an applied electric field and an applied magnetic field is straightforward. The primary difference in the results obtained is that with a magnetic field present, the mobility and the diffusion coefficient must be defined as tensor quantities.

PROBLEMS

14-1 Show that

$$\left(\frac{\mathbf{u}}{\nu_R}\cdot\nabla_r f_0 + \frac{\mathbf{F}}{m\nu_R}\cdot\nabla_u f_0\right)$$

is an odd function of $\mathbf{u}$ if $\mathbf{F} = q(\mathbf{E} + \mathbf{u}\times\mathbf{B})$.

14-2 Justify the small-velocity approximation used in Sec. 14-2.

14-3 Find the third approximation to the anisotropic distribution function discussed in Sec. 14-3. Find the nth approximation if ν_R = constant.

14-4 Show that if ν_R = constant and f_0 is a modified Maxwell-Boltzmann distribution function,

$$f_0(\mathbf{r},u) = n(\mathbf{r})\left[\frac{m}{2\pi\kappa T(\mathbf{r})}\right]^{3/2}\exp\left[\frac{-mu^2}{2\kappa T(\mathbf{r})}\right]$$

The heat flow equation (14-70) can be written

$$\mathbf{Q} = \mathcal{K}(n,T)\nabla T(\mathbf{r}) + \mathcal{L}(T)\nabla n(\mathbf{r}) + \mathcal{J}(n,T)\mathbf{E}$$

The factor $\mathcal{K}$ is called the "thermal conductivity."

14-5 Consider the Boltzmann equation with the collision term given by Eq. (14-15). Multiply each term in the Boltzmann equation by $m\mathbf{u}$ and integrate over all $\mathbf{u}$ (see Appendix A). Show that if ν_R is a constant, the relaxation collision frequency is equal to the defined effective collision frequency.

14-6 Can a diffusion coefficient or a mobility be defined if the collision frequencies are a function of position coordinates?

14-7 A system of plasma particles which moves only by diffusion has a uniform density at $t = 0$. At $t = 0^+$ and for all time, $n = n_0$ at $x = 0$ and $n = 0$ at $x = d$. Solve for the density of the particles between $x = 0$ and $x = d$ as a function of position and time. Find an approximate time for the decay from the initial to the final density distribution. List the assumptions and approximations made in obtaining a solution.

BIBLIOGRAPHY

Ferraro, V. C. A., and C. Plumpton: "An Introduction to Magneto-Fluid Mechanics," Oxford University Press, Fair Lawn, N.J., 1961.
sec. 7-10—The existence of a time of relaxation. On the assumed form of the collision term given in Eq. (14-15).

Rose, D. J., and M. Clark, Jr.: "Plasmas and Controlled Fusion," John Wiley & Sons, Inc., New York, 1961.
chap. 4, sec. 4—A discussion of mobility and diffusion very similar to that given in Secs. 14-3 to 14-5.
chap. 7, secs. 6 and 7—A discussion of mobility and diffusion in highly ionized plasmas.

chapter

15

The Pinch Effect

15-1 Introduction

Magnetic fields are caused by the motion of charged particles. If a charged particle moves with a constant velocity, the magnetic flux density set up by the particle motion is

$$\mathbf{B} = \frac{\mu_0 q \mathbf{u} \times \mathbf{R}}{4\pi |\mathbf{R}|^3} \tag{15-1}$$

where $\mathbf{R}$ is a vector from the charged particle to the field point. If we consider a positively charged particle moving in the z direction along the z axis, the magnetic field is only in the cylindrical coordinate θ direction, as shown in Fig. 15-1. If another similarly charged particle is allowed to move in the z direction parallel to the z axis, it will encounter a force $\mathbf{F} = q\mathbf{u} \times \mathbf{B} = -qu_z B_\theta \mathbf{a}_r$ because of its interaction with the magnetic field of the original moving particle. If we consider only magnetic effects, the particle moving off the z axis will be forced to move toward the z axis. Similarly, the particle originally in motion along the z axis will be forced off the z axis toward the other moving charged particle. The effect of the magnetic interaction between the two charged particles is to alter their motion in such a way that they are forced or pinched together.

Let us consider an infinitely long cylinder of plasma whose axis coincides with the z axis of the cylindrical coordinate system. If an electric field in the z direction is applied to the plasma, the electrons will move in the negative z direction and the ions will move in the positive z direction. The total magnetic field from both groups of charged particles will be in the positive θ direction. The force on all charged particles will be such as to cause the particles to move toward the z axis. If this magnetic force is sufficient to overcome the effects of the kinetic pressure of the plasma particles, the cylinder of plasma

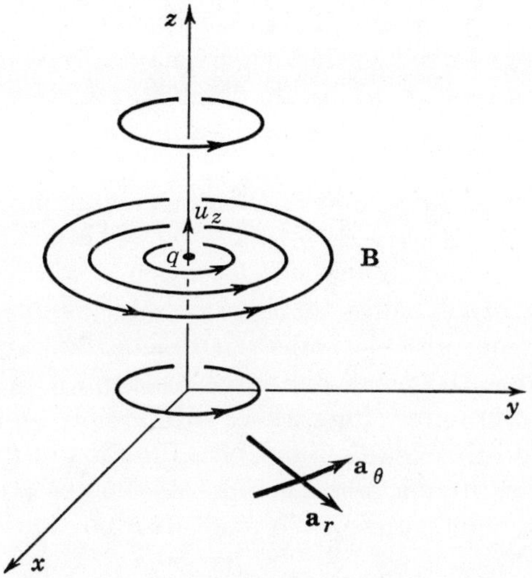

Fig. 15-1 The magnetic field of a positively charged particle moving along the z axis.

particles will contract or pinch because of its own magnetic field. This self-constriction of the plasma is known as the "pinch effect."

In the absence of instabilities, the pinch effect can be used to contain a high-energy plasma; this is the goal of the controlled-fusion research. The pinch effect may also play an important role in determining the behavior of the lightning discharge. Further, the pinch effect is important in that a comparison of the results of pinch experiments with the predictions of pinch theory reveals a good deal of information about the plasma state of matter. In this chapter we shall discuss two aspects of the pinch effect: (1) the quasi-equilibrium pinch, in which the radius of the plasma cylinder remains constant with time because of the balance between particle pressures and electromagnetic forces, and (2) the dynamic pinch, in which the radius of the plasma column decreases with time. In the next chapter we shall discuss the instabilities associated with the pinch effect.

15-2 The Quasi-equilibrium Pinch

We shall consider now the case in which the radius of an infinitely long plasma cylinder remains constant because of the balance of pressure and electromagnetic forces. This case is more general than it may appear. The time-varying pinch effect may in some cases be treated as a succession of quasi-equilibrium pinch states. In each succeeding state the current flowing is assumed to be increased by a small but finite amount.

Since we are considering a system in quasi-equilibrium, no parameters of the system will vary with time. The charged plasma particles under the influence of the applied electric field will move only in the positive and negative z directions. We shall assume that the drift velocities of the ions and electrons remain constant across the plasma cylinder. Because the pinch discharge is assumed to be cylindrically symmetrical and infinitely long, there will be only an azimuthal magnetic field B_θ, and this field will not be a function of z or of θ. It will be necessary to assume a relationship between the particle pressures, densities, and temperatures. We shall assume the ideal gas law to be valid.

$$P_i = n_i \kappa T_i \tag{15-2}$$
$$P_e = n_e \kappa T_e \tag{15-3}$$

where T_i and T_e are considered constant throughout the discharge. We shall further assume that $n_i = n_e = n$, $Z_i = 1$.

If we consider a *fully ionized* plasma, the behavior of the plasma is described by Eq. (8-4).

$$n_i m_i \frac{d\mathbf{U}}{dt} = \mathbf{j} \times \mathbf{B} - \nabla P \tag{8-4}$$

where $\mathbf{U}$ is the velocity of mass flow and $P = P_i + P_e$. We can expand $d\mathbf{U}/dt$ as follows:

$$\frac{d\mathbf{U}}{dt} = \frac{\partial \mathbf{U}}{\partial t} + (\mathbf{U} \cdot \nabla)\mathbf{U} \tag{15-4}$$

The first term on the right of Eq. (15-4) is zero because no parameters of the plasma vary with time. The second term on the right of Eq. (15-4) is zero because the mass flow is in the z direction and, since the plasma is infinite in the z direction, there can be no gradients of mass flow in that direction. Equation (8-4) can therefore be written

$$\mathbf{j} \times \mathbf{B} - \nabla P = 0 \tag{15-5}$$

where

$$\mathbf{j} = n|e|(\mathbf{v}_i - \mathbf{v}_e) \tag{15-6}$$

Note that Eq. (15-5) is identical with Eq. (8-5). Note also that Eq. (15-5) can be obtained directly from Eq. (8-1). Since $\mathbf{v}_e$ and $\mathbf{v}_i$ are caused by the same applied electric field, the electron velocity will be greater than the ion velocity. We can therefore approximate Eq. (15-6) as

$$\mathbf{j} = -n|e|\mathbf{v}_e \tag{15-7}$$

The substitution of Eqs. (15-2), (15-3), and (15-7) into Eq. (15-5) yields

$$n|e|\mathbf{v}_e \times \mathbf{B} + \kappa(T_i + T_e)\nabla n = 0 \tag{15-8}$$

Since there is only a B_θ and since the electron velocity is only in the negative z direction, Eq. (15-8) can be rewritten

$$n|e|v_e B_\theta + \kappa(T_i + T_e)\frac{dn}{dr} = 0 \tag{15-9}$$

where the total derivative is used because n is only a function of r. If we multiply Eq. (15-9) by r/n and then take the total derivative with respect to r, Eq. (15-9) becomes

$$\frac{d}{dr}\left(\frac{r}{n}\frac{dn}{dr}\right) + \frac{|e|v_e}{\kappa(T_i + T_e)}\frac{d}{dr}(rB_\theta) = 0 \tag{15-10}$$

We shall now make use of Maxwell's equation (4-2a) to eliminate B_θ from Eq. (15-10). In the present case Eq. (4-2a) can be written

$$\frac{1}{r}\frac{d}{dr}(rB_\theta) = \mu_0 n|e|v_e \tag{15-11}$$

where we have assumed that the plasma has the permeability of free space μ_0. Equations (15-10) and (15-11) can be combined to yield

$$\frac{d}{dr}\left(\frac{r}{n}\frac{dn}{dr}\right) + \frac{\mu_0|e|^2v_e^2}{\kappa(T_i + T_e)}nr = 0 \tag{15-12}$$

which is the equation for the variation of n as a function of r.

The particle densities must be symmetrical about the z axis. It therefore follows that $dn/dr = 0$ when $r = 0$. With this boundary condition, the solution to Eq. (15-12) is

$$n = \frac{n_0}{(1 + n_0br^2)^2} \tag{15-13}$$

where

$$b \equiv \frac{\mu_0|e|^2v_e^2}{8\kappa(T_e + T_i)}$$

and n_0 is the particle density at $r = 0$. The variation of the particle density as a function of the distance from the z axis as given by Eq. (15-13) is known as the "Bennett distribution." A plot of the Bennett distribution is shown in Fig. 15-2. The distribution falls off very sharply with r, and therefore the plasma is concentrated in the region of the z axis. Although the Bennett distribution predicts a particle density extending to infinity, for all practical purposes the particle density falls to zero at some radius which we shall call r_p.

We shall now calculate an approximate relationship between the current which flows in the pinch discharge, the particle densities, and the particle temperatures. The total number of electrons (or ions) per unit length of the discharge is given by

$$N_l = \int_0^{r_p} 2\pi rn\, dr \tag{15-14}$$

If Eq. (15-13) is substituted into Eq. (15-14), the resulting equation is

$$N_l = \int_0^{r_p} \frac{2\pi rn_0\, dr}{(1 + n_0br^2)^2} \tag{15-15}$$

which when integrated becomes

$$N_l = \frac{n_0\pi r_p^2}{1 + n_0br_p^2} \tag{15-16}$$

Equation (15-16) can be solved for n_0.

$$n_0 = \frac{N_l}{r_p^2(\pi - N_lb)} \tag{15-17}$$

Fig. 15-2 The Bennett distribution.

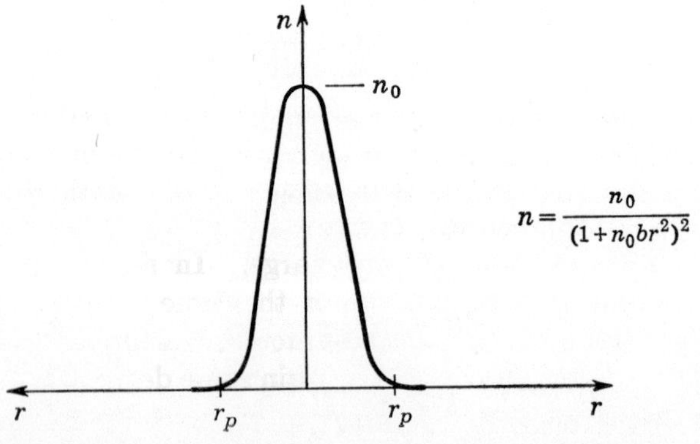

If n_0 from Eq. (15-17) is substituted for n_0 in the denominator of Eq. (15-13), Eq. (15-13) becomes

$$n = \frac{n_0}{[1 + br^2N_l/r_p^2(\pi - N_lb)]^2} \tag{15-18}$$

Since at $r = r_p$ the particle density is essentially zero, Eq. (15-18) for $r = r_p$ can be written

$$0 = \frac{n_0}{1 + bN_l/(\pi - N_lb)} \tag{15-19}$$

It follows that $(\pi - N_lb) = 0$ or

$$N_lb = \pi \tag{15-20}$$

If Eq. (15-20) is combined with the definition of b so as to eliminate b, we find that

$$(N_l|e|v_e)^2 = \frac{8\pi N_l \kappa(T_e + T_i)}{\mu_0} \tag{15-21}$$

The current in the discharge, neglecting ion current, is

$$I = N_l|e|v_e \tag{15-22}$$

From a comparison of Eq. (15-21) with Eq. (15-22), it is apparent that

$$I^2 = \frac{8\pi N_l \kappa(T_e + T_i)}{\mu_0} \tag{15-23}$$

Equation (15-23) is known as the "Bennett relation." In deriving the Bennett relation we have assumed that $\mathbf{v}_e$ is not a function of radial position. It can, however, be shown that Eq. (15-23) is valid independent of the assumed form of $\mathbf{v}_e(r)$. It appears from Eq. (15-23) that a plasma can be contained at any temperature by allowing the appropriate current to flow through the pinched plasma. The temperature attainable in a pinched discharge is, however, limited by the bremsstrahlung losses from the discharge. In addition, the occurrence of numerous instabilities limits the time that a high-energy plasma can be contained. In actual practice it is necessary to employ a magnetic field in the z direction to obtain some degree of stabilization in the pinch discharge.

15-3 The Dynamic Pinch

The relations we have developed for the quasi-equilibrium pinch are applicable only for those systems in which the pinch current is slowly varying compared with the time it takes for the plasma particles to attain constant temperatures. In some pinch discharges, the current change and the pinch time are so rapid that the plasma particles cannot attain thermal equilibrium. It is the behavior of this dynamic pinch which we shall now investigate.

We shall analyze the dynamic pinch effect using the "snowplow" model. We shall assume the fully ionized plasma to be a perfect conductor confined at time $t = 0^-$ by a cylindrical dielectric vessel of radius r_0. We shall again assume that the plasma can be considered infinite in the axial direction. At $t = 0^+$ an electric field is applied to the plasma in the direction of the cylindrical axis. The current which flows through the pinch discharge will be determined by the inductance, resistance, and capacitance of both the external circuit which applies the electric field to the plasma cylinder and the discharge itself.

Since the plasma is assumed to be a perfect conductor, the applied electric field and the resultant plasma current and magnetic field cannot penetrate the plasma. At $t = 0^+$, the plasma current will flow in an infinitely thin cylindrical shell of radius r_0. There will be no current, electric field, or magnetic field within the plasma cylinder. The magnetic flux density at $r = r_0$ (just outside the current layer at $t = 0^+$) can be found from Eq. (4-2b).

$$B_\theta\Big|_{r=r_0} = \frac{\mu_0 I\Big|_{t=0^+}}{2\pi r_0} \qquad (15\text{-}24)$$

where $\mathbf{B} = \mu_0\mathbf{H}$. The magnetic pressure inward is

$$P_m = \frac{|\mathbf{B}|^2}{2\mu_0} \qquad (8\text{-}11)$$

evaluated at $r = r_0$. The total inward force per length on the current shell at $t = 0^+$ due to B_θ is $-(2\pi r P_m)\mathbf{a}_r\Big|_{r=r_0}$

or

$$\mathbf{F}_l = -\frac{\mu_0 I^2\Big|_{t=0^+}}{4\pi r_0}\,\mathbf{a}_r \qquad (15\text{-}25)$$

We shall assume that as the cylindrical current shell moves inward because of the pinch effect, the cylindrical shell acts like a "magnetic piston" and sweeps up all of the charged particles it encounters (hence the name "snowplow"). If we ignore the effects of kinetic pressure, we can set the magnetic force on the current shell equal to the rate of change of momentum of the current shell for any radius r of the cylindrical shell.

$$\frac{d}{dt}\left(M_l \frac{dr}{dt}\right) = -\frac{\mu_0 I^2(t)}{4\pi r} \tag{15-26}$$

where M_l is the mass per unit length of the particles swept up by the cylindrical shell when it has reached a radius r. If ρ_M is the original mass density of the plasma, the mass per unit length which sticks to the cylindrical shell is

$$M_l = \pi \rho_M (r_0^2 - r^2) \tag{15-27}$$

Equations (15-26) and (15-27) can be combined to yield

$$\frac{d}{dt}\left[\pi \rho_M (r_0^2 - r^2) \frac{dr}{dt}\right] = -\frac{\mu_0 I^2}{4\pi r} \tag{15-28}$$

Equation (15-28) describes the variation of the radius of the dynamic pinch discharge as a function of time and of the pinch current.

Equation (15-28) can be solved numerically if an expression for the pinch current can be obtained. The pinch current is determined by the parameters of the pinch discharge itself and by the external electrical circuit which supplies the energy for the discharge. In order to create the pinch current, it is usual to store charge in capacitor banks and then to discharge those banks across a spark gap. Thus the external circuit which supplies the energy for the pinch discharge usually has associated with it a large capacitance. If the electrical circuit external to the pinch discharge has a greater inductance than the pinch itself, the pinch current will be controlled primarily by the external circuit. Under these conditions the current will be a damped sinusoid or the initial portion of a damped sinusoid. The frequency of the pinch current and the amount of damping (usually negligible) will be determined by the resistance, inductance, and capacitance of the external electrical circuit. If the pinch discharge itself has a greater inductance than the external electrical circuit, the pinch discharge plays an important role in determining the pinch current. The inductance of the pinch discharge will, in general, be a function of the radius of the pinch cylinder. We can write an expression for the voltage drop across the pinch discharge

$$V = \frac{d}{dt}[L(r)I] \tag{15-29}$$

where $L(r)$ is the inductance of the pinch column. Equation (15-29) can be combined with Eq. (15-28) and with the circuit equations to provide a set of equations which can be solved for the pinch radius as a function of time.

In using the snowplow model, we have treated the pinch discharge as if it were a column of fluid whose electrical conductivity was infinite and whose kinetic pressure was zero. It is possible to modify the simple dynamic theory which we have considered in order to take account of the effects of a finite plasma kinetic pressure and a finite plasma conductivity.

The dynamic pinch, like the quasi-equilibrium pinch, is subject to a large number of instabilities. These instabilities limit the lifetime of the dynamic pinch discharge.

PROBLEMS

15-1 Show that for the quasi-equilibrium pinch Eq. (15-5) follows directly from Eq. (8-1).

15-2 Plot I versus T for the quasi-equilibrium pinch discharge if the pinch diameter is 1 cm and if $n_e = 10^{16}$ electrons per cubic meter. Assume $T_e = T_i = T$, and consider temperatures up to 10^7°K (10^3 ev). What is the bremsstrahlung loss per meter of the pinch column if $T = 10^7$°K?

15-3 Show that the magnetic field inside the quasi-equilibrium pinch column is

$$B_\theta = \frac{\mu_0 |e| v_e n_0 r}{2(1 + n_0 b r^2)}$$

15-4 We have noted that Eq. (15-5) is identical with Eq. (8-5). In Chap. 8 we showed that the relation

$$\nabla\left(P + \frac{|\mathbf{B}|^2}{2\mu_0}\right) = \frac{(\mathbf{B}\cdot\nabla)\mathbf{B}}{\mu_0} \tag{8-8}$$

follows from Eq. (8-5). By direct substitution prove that Eq. (8-8) is valid for the quasi-equilibrium pinch discharge.

15-5 Using the techniques developed in Prob. 8-4, find an expression for the radial electric field present in the quasi-equilibrium pinch discharge. Assume that collisions are negligible. Discuss.

15-6 In Chap. 8 we developed the concept of a magnetic pressure for the case of straight magnetic field lines. Justify the use of this magnetic pressure [Eq. (8-11)] in the discussion of the dynamic pinch.

15-7 An ideal voltage source V_0 is impressed across a neutral gas with the result that a dynamic pinch discharge is initiated. Assume that the inductance of the pinch discharge is

$$L = \frac{\mu_0}{2\pi} \ln \frac{r_a}{r}$$

where $r < r_a$. Show that the equation which specifies the behavior of the pinch radius is

$$\frac{d}{dt}\left[\rho_M(r_0^2 - r^2)\frac{dr}{dt}\right] = -\frac{\left(\int_0^t V_0\,dt\right)^2}{\mu_0 r\left(\ln \frac{r_a}{r}\right)^2}$$

15-8 In discussing the pinch effect, we have assumed that the pinch current is in the z direction and that the pinch magnetic field is therefore in the θ direction. In the so-called "θ pinch," the pinch current is constrained to be in the θ direction, causing the pinch magnetic field to be in the z direction. Using the snowplow model, set up the equations of motion for the θ pinch.

15-9 Obtain an equation for the quasi-equilibrium pinch effect in which $\mathbf{v}_e(r)$ is an arbitrary function of radial position.

BIBLIOGRAPHY

Allis, W. P.: "Nuclear Fusion," D. Van Nostrand Company, Inc., Princeton, N.J., 1960.
chap. 10—Sustained Pinches.

Bennett, W. H.: Magnetically Self-focusing Streams, *Phys. Rev.*, **45**:890 (1934).
The original article on the theory of a constricted gas discharge.

Cole, G. N. A.: The Pinch Effect, *Sci. Progr. (London)*, **47**:437 (1959).
A general discussion of the pinch effect.

Evans, W. H.: Magnetic Pinch in Relation to the Lightning Stroke, *Geofis. Pura Appl.*, **52**:162 (1962).
A calculation of the pinch effect in a lightning stroke.

Glasstone, S., and R. H. Lovberg: "Controlled Thermonuclear Reactions," D. Van Nostrand Company, Inc., Princeton, N.J., 1960.
chap. 7—The Pinch Effect.

Rose, D. J., and M. Clark, Jr.: "Plasmas and Controlled Fusion," John Wiley & Sons, Inc., New York, 1961.
chap. 14—Devices Based upon the Pinch Effect. Includes a general derivation of the Bennett relation [Eq. (15-23)].

Rosenbluth, M. N., R. Garvin, and A. Rosenbluth: Infinite Conductivity Theory of the Pinch, Report LA-1850, Los Alamos Scientific Laboratory, September, 1954.
On the derivation and solution of Eq. (15-28) for the case that the inductance of the pinch discharge is greater than the inductance of the external electrical circuit.

chapter

16

Position-space Plasma Instabilities

16-1 Introduction

Interactions between electromagnetic fields and plasmas exist for which time-invariant solutions to the equations describing the interactions can be found. Under such conditions it is usual to say that the plasma is "confined" by the fields. We have analyzed one such confining interaction in our discussion of the quasi-equilibrium pinch effect. The existence of an equilibrium state does not, however, imply that that particular equilibrium state is stable. A rubber ball balanced on the top of a stick is in an unstable equilibrium state. Any slight displacement of the ball will result in the destruction of the equilibrium state. On the other hand, a rubber ball at the bottom of a valley is in a stable equilibrium. After any slight displacement of the ball, the ball will return to the bottom of the valley. In any equilibrium system the criterion for stability is as follows: If a small perturbation from the equilibrium state is followed by a return to equilibrium or by a nongrowing oscillation about the equilibrium state, the system is stable; if a small perturbation from the equilibrium state results in a growth of the perturbation, the system is unstable.

In this chapter we shall consider position-space plasma instabilities, instabilities which are dependent on the structure of the plasma in real space. In Chap. 17 we shall discuss velocity-space plasma instabilities, instabilities which are dependent upon the form of the electron and ion velocity distribution functions.

16-2 Instabilities

We shall consider now one of the simplest equilibrium configurations, a system in which the confinement of a perfectly diamagnetic plasma is effected by means of a static magnetic field. We shall assume that the plasma has a uniform kinetic pressure P_0. If the perfectly diamagnetic plasma is to be confined by a static magnetic field, the magnetic pressure at the plasma surface must be equal to the kinetic pressure of the plasma.

$$P_0 = \frac{|\mathbf{B}_0|^2}{2\mu_0} \tag{16-1}$$

where $\mathbf{B}_0$ is the magnetic field at the plasma surface. Suppose the confining field lines are curved so that they are concave toward the plasma as shown in Fig. 16-1. The magnetic field strength diminishes with increasing distance from the center of curvature of the field lines. Thus in equilibrium, $\mathbf{B}_0$ is the largest magnetic field present in the system.

In order to test this equilibrium system for stability, we must perturb the plasma surface and then determine whether the perturbations will grow larger or diminish in size. We shall perturb the plasma surface with a wave whose crests and troughs run parallel to the magnetic field lines as shown in Fig. 16-2. The kinetic pressure of the plasma will be unchanged if the plasma is perturbed in such a manner that its volume does not change. The magnetic field along the plasma surface will, however, be changed. The magnetic field in the troughs will be greater than B_0; the magnetic field on the crests will be less than B_0. In the troughs, therefore, the magnetic pressure will exceed the kinetic pressure; on the crests the kinetic pressure will exceed the magnetic pressure. The troughs will, therefore, become deeper and

Fig. 16-1 A perfectly diamagnetic plasma contained by a magnetic field concave toward the plasma.

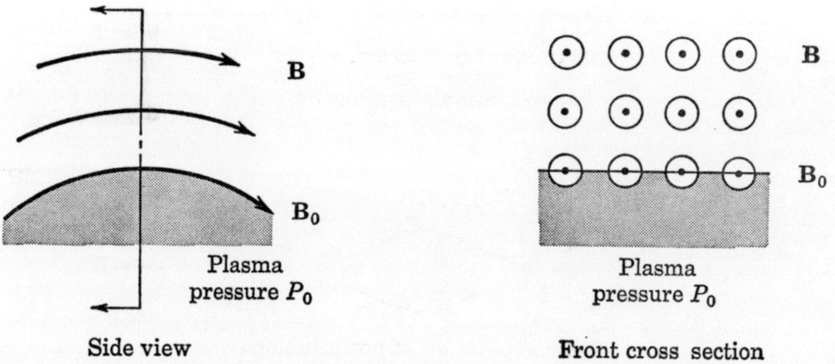

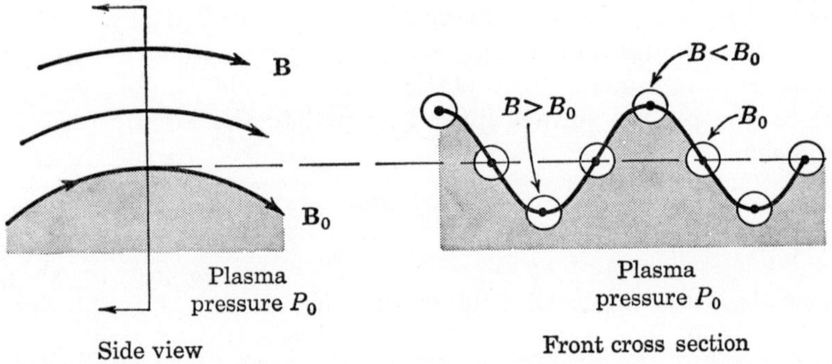

Fig. 16-2 The plasma-field configuration shown in Fig. 16-1 after the plasma surface has been perturbed.

the crests higher. The initial perturbation will grow. The equilibrium system is unstable.

As an example of the foregoing analysis, let us examine the pinch effect. If we consider the equilibrium column of pinched plasma to be perfectly diamagnetic, our foregoing physical analysis can be easily applied. We shall perturb the surface of the equilibrium pinch discharge in the manner shown in Fig. 16-3.

The perturbations along the surface of the pinch column are assumed to be applied in such a way that the kinetic pressure of the plasma is

Fig. 16-3 A column of pinched plasma before and after application of cylindrically symmetrical perturbations.

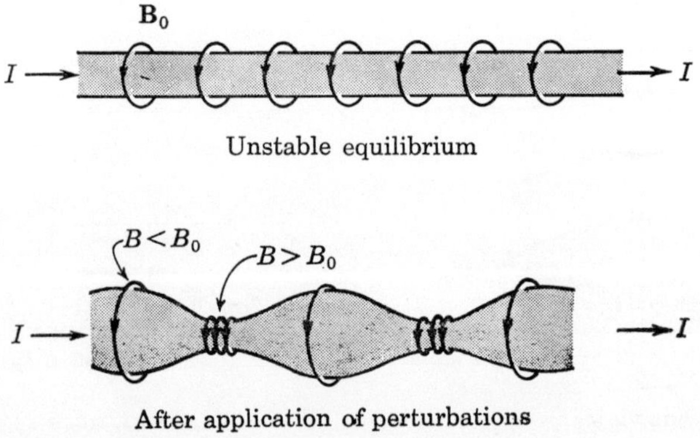

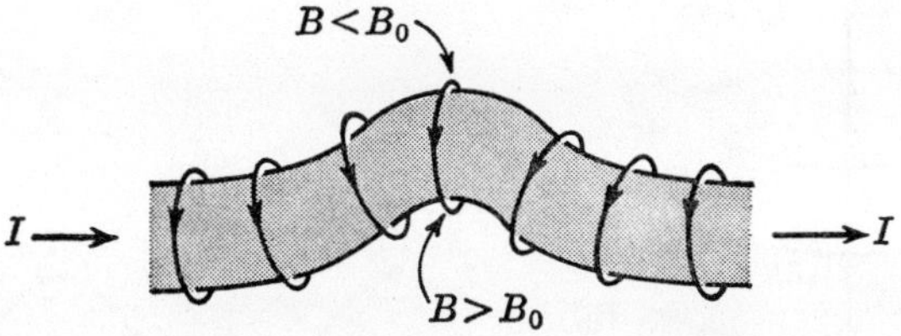

Fig. 16-4 The kink instability.

not changed. Since the azimuthal magnetic field at the surface of the plasma is inversely proportional to the radius of the plasma [from Eq. (4-2*b*)], the regions of decreased radius will suffer an increased magnetic pressure. Similarly, the regions of increased radius will suffer a decreased magnetic pressure. Thus the crests of the perturbation will grow in size, and the troughs of the perturbation will become further constricted. The initially straight plasma cylinder distorts into what looks like a string of sausages. This type of pinch instability is called the "sausage" instability. The sausage instability is one of the infinite number of pinch instabilities. The instabilities other than the sausage are not cylindrically symmetrical, as is the sausage instability. Another pinch instability, the so-called "kink" instability, is shown in Fig. 16-4. The magnetic field strength on the underside of the pinch column shown in Fig. 16-4 is greater than the magnetic field strength on the top side. Thus the kink will increase in size. The time that a plasma can be contained by the pinch effect is limited

Fig. 16-5 A perfectly diamagnetic plasma contained by a magnetic field convex toward the plasma.

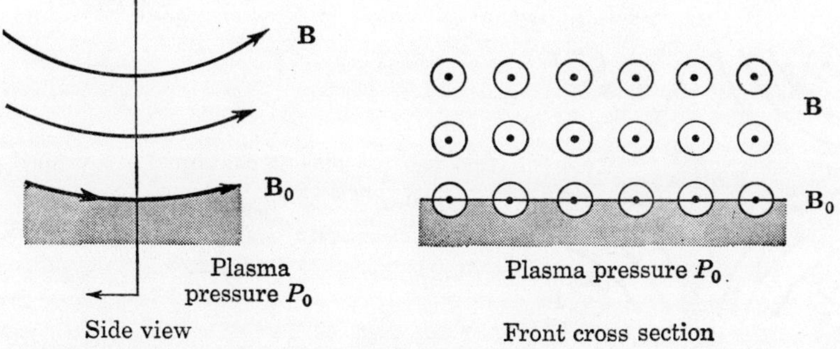

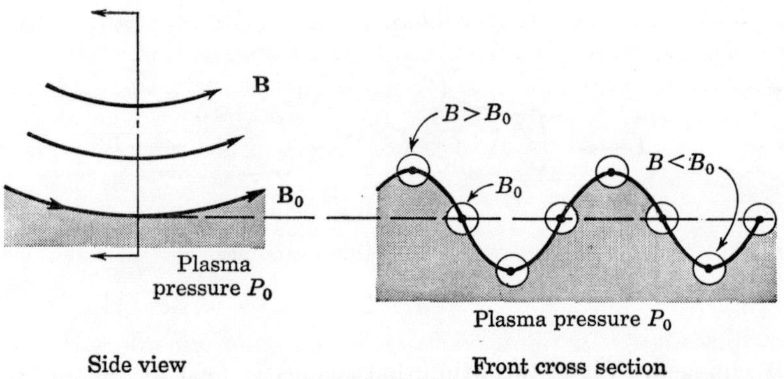

Fig. 16-6 The plasma-field configuration shown in Fig. 16-5 after the plasma surface has been perturbed.

by the appearance of the sausage, kink, and other pinch instabilities. In practice the pinched plasma is not perfectly diamagnetic, and very often fields other than those we have considered are present. The calculation of the pinch system stability becomes, in general, a difficult task.

We have investigated the stability of a perfectly diamagnetic plasma confined by a static magnetic field which is concave toward the plasma. We shall consider now the case of an equilibrium plasma confined by a magnetic field which is convex toward the plasma as shown in Fig. 16-5. The magnetic field strength increases in a direction away from the plasma. The effect of perturbing the plasma surface is shown in

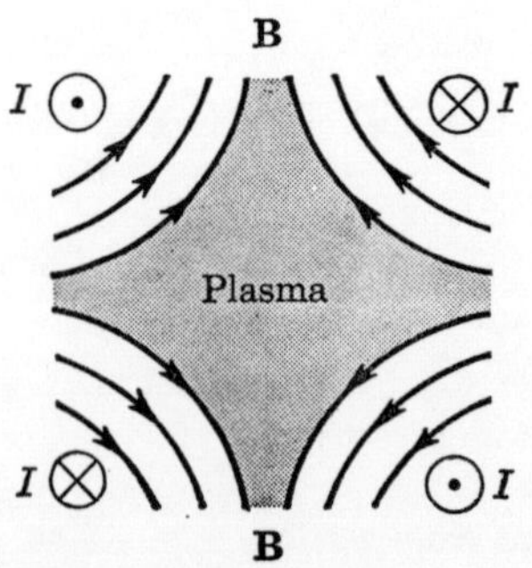

Fig. 16-7 A plasma contained in a cusped magnetic field.

Fig. 16-6. The magnetic pressure on the crests is greater than the kinetic pressure; the kinetic pressure in the troughs is greater than the magnetic pressure in the troughs. We have again assumed that the kinetic pressure is unaffected by the perturbations. The plasma is forced to return to its equilibrium configuration. The equilibrium system is stable.

We may conclude that a perfectly diamagnetic plasma can be confined in a stable manner if the confining magnetic field is everywhere convex toward the plasma. Such a magnetic field is commonly referred to as "cusped." A simple example of a cusped geometry is shown in Fig. 16-7. The cusped field configuration can be produced by four current-carrying wires as shown in Fig. 16-7.

16-3 Discussion

We have tested for plasma stability by assuming a wavelike perturbation of the plasma surface and noting the behavior of those perturbations. This procedure is somewhat artificial, since in an actual equilibrium system the perturbations will usually be of a random nature. It follows, though, that if a system is unstable to wavelike perturbations, it will, in general, be unstable to random perturbations.

The foregoing qualitative discussion of plasma stability has been restricted to the consideration of a perfectly diamagnetic plasma. It is not clear physically what effects a magnetic field within the plasma will have on the stability of the plasma. Further, our qualitative discussion is incapable of predicting the rates of growth of various instabilities. For a more complete understanding of plasma stability, it is necessary to resort to the mathematics of plasma physics. The mathematics involved in the investigation of plasma stability is complex. Suffice it to say that plasma stability can be studied from either a hydrodynamic or a Boltzmann equation point of view. Both approaches are used. In either case it is necessary to make many approximations to arrive at a tractable set of equations. The theories of plasma stability have much in common with the theories of plasma waves since instabilities can be treated as plasma waves which grow in amplitude.

PROBLEMS

16-1 Consider the two current-carrying wire loops in Fig. P16-1. Show that if currents flow around the two loops in the same direction, a magnetic mirror field is created between the loops. Show that if currents flow around the two loops in opposite directions, a cusped magnetic field is created between the loops.

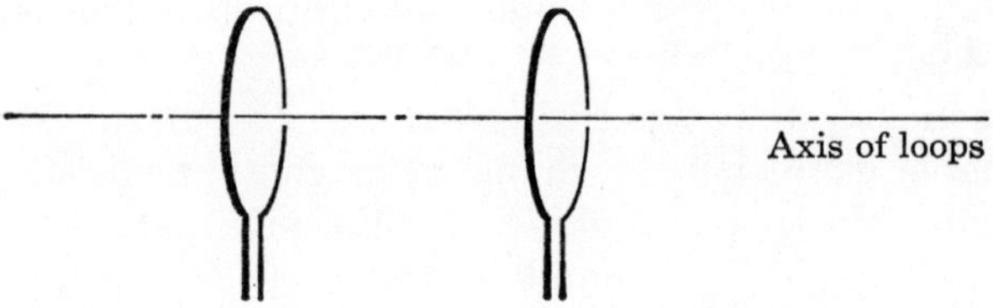

Fig. P16-1 Two current-carrying loops.

16-2 Can the motion of a charged particle in the cusped magnetic field found in Prob. 16-1 be analyzed in terms of perturbation drifts? What is the magnitude of the magnetic field on the axis midway between the two loops?

16-3 A magnetic mirror configuration is stable near the ends of the mirror and unstable near the middle of the mirror. Explain.

16-4 The magnetic field of a toroid (see Prob. 5-8 and Fig. P5-8) is unstable at its outer periphery and stable at its inner periphery. Explain.

16-5 One disadvantage of using the cusped magnetic field geometry for containing a high-temperature plasma is that there are certain avenues of escape available for the charged plasma particles. How can the charged particles escape from the cusped field shown in Fig. 16-7? The cusped field found in Prob. 16-1?

BIBLIOGRAPHY

Glasstone, S., and R. H. Lovberg: "Controlled Thermonuclear Reactions," D. Van Nostrand Company, Inc., Princeton, N.J., 1960.
chap. 13—Plasma Stability Theory.

Rose, D. J., and M. Clark, Jr.: "Plasmas and Controlled Fusion," John Wiley & Sons, Inc., New York, 1961.
chap. 12—Plasma Stability.

chapter

17

Velocity-space Plasma Instabilities

17-1 Introduction

In the preceding chapter we discussed plasma instabilities which arise in certain plasma-field configurations. In this chapter we shall consider plasma instabilities whose characteristics are dependent upon the form of the plasma particle phase-space distribution function. Since the distribution functions are velocity-dependent, we shall classify these types of instabilities as "velocity-space" instabilities. We shall generalize our definition of instability to include *both* growing waves and decaying waves.

It is convenient in discussing wave propagation to consider only plane-wave solutions to the plasma equations. This approach is sufficiently general, for any solution can be found by the appropriate combination of plane waves. The spatial and temporal variation of a one-dimensional plane wave is $\exp[i(k_x x - \omega t)]$. In Chap. 7 we have assumed that the angular frequency ω is a real quantity. It is, however, possible to find solutions to the propagation equations for which ω is a complex quantity

$$\omega = \omega_r + i\gamma \tag{17-1}$$

Under these conditions the spatial and temporal variation of the wave becomes

$$\exp(ik_x x - i\omega_r t + \gamma t) \tag{17-2}$$

The propagation exponent contains a real quantity γt. If $\gamma > 0$, the wave will increase in amplitude as time increases. If $\gamma < 0$, the wave will decrease in amplitude as time increases. This type of wave behavior is not to be confused with the spatial damping of an electromagnetic wave due to a complex propagation constant. The growing and decaying waves we shall discuss in this chapter have amplitudes which are temporally, not spatially, dependent.

We shall discuss in a semiquantitative manner the occurrence of temporally growing and decaying *longitudinal* waves in collisionless plasmas. The temporal damping of the longitudinal electron oscillations in a plasma is known as Landau damping. We mentioned the occurrence of Landau damping in Chap. 11. No general name is given to temporally damped longitudinal ion oscillations or to temporally growing longitudinal plasma oscillations.

17-2 Growing and Decaying Longitudinal Plasma Oscillations

If longitudinal oscillations are present in a plasma, there will be a certain amount of energy associated with the organized motion of the oscillating particles. In addition, there will be energy associated with the disorganized thermal motion of the particles. We shall consider the case in which the thermal energy is much greater than the organized energy of the oscillating particles. If the oscillations remain constant in amplitude, the division of energy between organized and unorganized motion remains constant. If the amplitude of the longitudinal oscillations grows, there must be an increase in the energy of organized motion at the expense of the thermal energy. If the longitudinal oscillations are damped, there must be an increase in thermal energy at the expense of the energy of organized motion. As we shall see, the mechanism of the energy transfer which yields growing and decaying waves is the trapping of the charged particles by the electric field of the oscillations into the moving potential troughs of the wave.

We shall assume that a longitudinal wave is established in a plasma at $t = 0$ whose electric field is given by $E_x(x,t) = E_0 \sin (k_x x - \omega_r t)$. The phase velocity of the wave is $v_p = \omega_r/k_x$. Charged particles moving with velocities approximately equal to the phase velocity of the wave can be trapped in a stable position by the wave and carried along by it. Particles moving with a velocity approximately equal to v_p will, in the coordinate system moving along with the wave, experience a time-independent force $F_x(x) = qE_0 \sin k_x x$. If the particle displacement is small, $\sin k_x x \cong k_x x$ and the equation of motion for the particles is

$$m \frac{d^2x}{dt^2} = qE_0 k_x x \tag{17-3}$$

The particles will thus begin to oscillate with an angular frequency

$$\omega_0 = \left(\frac{qE_0k_x}{m}\right)^{1/2} \tag{17-4}$$

If we consider the one-dimensional case, the phase-space distribution function for the particles can be expressed solely as a function of u_x and x. We shall assume a uniform density of particles in position coordinates at $t = 0$. Thus we shall write the "one-dimensional" particle distribution function at $t = 0$ as $f_0(u_x)$. All particles whose x components of velocity are in the range

$$(v_p - \Delta u_x) \leq u_x \leq (v_p + \Delta u_x)$$

will be caught by the wave and carried along with it if Δu_x corresponds to the depth of the potential well in which the particles are trapped.

The particles will undergo a change in energy in the process of being trapped. Before trapping, each particle of velocity u_x has an energy $mu_x^2/2$. After trapping, the energy of the same particle is $m[v_p^2 + (u_x - v_p)^2]/2$, where $mv_p^2/2$ is the particle energy in the absence of oscillation and $(m/2)(u_x - v_p)^2$ is the particle energy of oscillation. The energy change due to trapping of a particle whose x component of velocity was originally u_x is

$$\Delta \mathcal{E} = \frac{m}{2}\,[v_p^2 + (u_x - v_p)^2 - u_x^2] \tag{17-5}$$

The total energy change of the particles per unit volume due to trapping is therefore

$$\Delta W = \frac{m}{2}\int_{v_p-\Delta u_x}^{v_p+\Delta u_x} f_0(u_x)[v_p^2 + (u_x - v_p)^2 - u_x^2]\,du_x \tag{17-6}$$

If we expand $f_0(u_x)$ in a power series about v_p and retain only the first two terms in the series, the distribution function can be written

$$f_0(u_x) = f_0(v_p) + \left.\frac{\partial f_0}{\partial u_x}\right|_{u_x=v_p}(u_x - v_p) \tag{17-7}$$

After substitution of Eq. (17-7) into Eq. (17-6) and expansion of Eq. (17-6), we find that

$$\Delta W = -\frac{m}{2}\int_{v_p-\Delta u_x}^{v_p+\Delta u_x} f_0(v_p)2v_p(u_x - v_p)\,du_x - \frac{m}{2}\int_{v_p-\Delta u_x}^{v_p+\Delta u_x}\left.\frac{\partial f_0}{\partial u_x}\right|_{u_x=v_p} 2v_p(u_x - v_p)^2\,du_x \tag{17-8}$$

After Eq. (17-8) is integrated and evaluated at the limits, the expression for the energy change of the particles per unit volume due to trapping becomes

$$\Delta W = -\tfrac{2}{3} m v_p (\Delta u_x)^3 \left.\frac{\partial f_0}{\partial u_x}\right|_{u_x = v_p} \tag{17-9}$$

Equation (17-9) is physically plausible. If there are more particles moving slightly faster than the phase velocity than there are moving slower, the derivative of the distribution function at $u_x = v_p$ is positive. Thus the trapping process will tend to slow down more particles than it speeds up, and there will be a net decrease in the thermal energy of the particles. Energy will be transferred from thermal energy to the energy of organized motion, and the amplitude of the longitudinal oscillation will grow. If, on the other hand, there are more particles moving slower than the phase velocity than there are moving faster, the derivative of the distribution function at $u_x = v_p$ is negative. The trapping process will speed up more particles than it slows down, and there will be a net increase in the thermal energy of the particles. Energy will be transferred from the energy of organized motion into thermal energy, and the longitudinal wave will be damped. For Maxwell-Boltzmann statistics the derivative of the distribution function is negative for all velocities. Special devices must be employed to create a distribution function which has a positive derivative. One example of such a distribution function is the distribution function associated with an electron beam. The primary velocity component present in an electron beam is that of the beam velocity. There are only a few particles in the beam which have velocities higher or lower than the beam velocity. The distribution function resembles a Gaussian curve with its peak at the beam velocity.

The energy-change expression [Eq. (17-9)] represents a change in the energy of thermal motion per unit volume of the particles. As we have discussed, this change must be equal and opposite to the change in the energy of the plasma oscillations. The change in particle energy takes place in approximately $\tau_t = 1/\omega_0$. Thus we can write an order-of-magnitude equation for the rate of change of the energy per unit volume associated with the plasma oscillations.

$$\frac{d\mathfrak{U}}{dt} = -\omega_0 \, \Delta W \tag{17-10}$$

where

$$\mathfrak{U} = \tfrac{1}{2}\epsilon_0 E_0^2 \tag{17-11}$$

Equations (17-10), (17-11), and (17-4) can be combined to yield

$$\frac{d\mathfrak{U}}{dt} = \frac{2}{3}\left(\frac{qE_0k_x}{m}\right)^{1/2} m v_p (\Delta u_x)^3 \left.\frac{\partial f_0}{\partial u_x}\right|_{u_x = v_p} \tag{17-12}$$

We shall now express Δu_x in terms of the electric field. The change in potential energy from crest to null in the wave is

$$-\int_{-\pi/2}^{0} qE_0 \sin k_x x \, dx = \frac{qE_0}{k_x}$$

In order for the particle to be trapped by the wave, therefore,

$$\frac{1}{2} m(\Delta u_x)^2 = \frac{qE_0}{k_x}$$

or

$$\Delta u_x = \left(\frac{2qE_0}{mk_x}\right)^{1/2} \tag{17-13}$$

If Eq. (17-13) is substituted into Eq. (17-12), the resultant equation is

$$\frac{d\mathfrak{U}}{dt} = \frac{4\sqrt{2}}{3} \frac{v_p}{mk_x} (qE_0)^2 \left.\frac{\partial f_0}{\partial u_x}\right|_{u_x = v_p} \tag{17-14}$$

Equation (17-14) can be expressed in terms of the energy density associated with the electric field.

$$\frac{d\mathfrak{U}}{dt} = \left[\frac{8\sqrt{2}}{3} \frac{\omega_r}{n_0} \left(\frac{\omega_p}{k_x}\right)^2 \left.\frac{\partial f_0}{\partial u_x}\right|_{u_x = v_p}\right] \mathfrak{U} \tag{17-15}$$

where $\omega_p^2 = q^2 n_0/m\epsilon_0$ and $v_p = \omega_r/k_x$. Again we see that if the slope of the distribution function at $u_x = v_p$ is positive, the plasma wave will grow. If the slope of the distribution function at $u_x = v_p$ is negative, the plasma wave will decay as time goes forward. For a field variation of $e^{\gamma t}$, the energy varies as the field squared or as $e^{2\gamma t}$. The coefficient of $\mathfrak{U}$ on the right-hand side of Eq. (17-15) is thus identifiable as 2γ, and therefore

$$\gamma = \frac{4\sqrt{2}}{3} \frac{\omega_r}{n_0} \left(\frac{\omega_p}{k_x}\right)^2 \left.\frac{\partial f_0}{\partial u_x}\right|_{u_x = v_p} \tag{17-16}$$

Equation (17-16) is an expression for the imaginary part of the complex angular frequency. It can be either positive (growing waves) or negative (decaying waves) depending on the sign of $\left.\partial f_0/\partial u_x\right|_{u_x = v_p}$.

17-3 Discussion

In the foregoing discussion we assumed that the longitudinal plasma oscillations have a small energy content compared with the thermal motion of the plasma. Thus it is a reasonable approximation to allow the distribution function $f_0(u_x)$ to remain constant as the oscillations grow or are damped. If the plasma oscillations extract a significant amount of energy from the thermal motion, the equilibrium distribution function will be distorted as time increases. Under these conditions our analysis can be considered only qualitative. The criterion for the validity of the approximation that the distribution function be time-invariant is that the trapping time $\tau_t = 1/\omega_0$ be very long compared with the growing or damping time $1/\gamma$.

We have discussed velocity-space instabilities in a semiquantitative manner. In this way it has been possible to obtain a physical understanding of the instability mechanism. Our treatment, however, has not been rigorous. Only under special conditions is Eq. (17-16) a good approximation to the imaginary part of the complex frequency derived by more rigorous techniques. In order to obtain information about all types of velocity-space instabilities, it is necessary to make use of the Boltzmann equation. The Boltzmann equation may be manipulated using techniques similar to those introduced in Chap. 11. It is necessary, however, to consider initial conditions in the concurrent solution of the Boltzmann equation and of Poisson's equation. Further, poles in the integrands of expressions derived from the Boltzmann equation and from Poisson's equation, such as the pole in the integrand of Eq. (11-22), cannot be ignored. It is these poles which provide the mathematical basis for velocity-space instabilities.

PROBLEMS

17-1 The criterion for the validity of the mathematical approach to velocity-space plasma instabilities used in this chapter is that the trapping time be very long compared with the growing or damping time. Reformulate this criterion in terms of the maximum electric field which may be present in the longitudinal oscillations.

17-2 How does the expression for the maximum electric field derived in Prob. 17-1 depend on the wavelength of the longitudinal wave? Discuss.

17-3 Using the maximum electric field found in Prob. 17-1, compare the maximum potential energy change from crest to null of the longitudinal wave with the average thermal energy of the particles if

$$k_x^2 = \frac{\omega_p^2}{\overline{u_x^2}} \qquad \text{and} \qquad \overline{u_x^2} = \frac{\kappa T}{m}$$

Discuss.

17-4 Calculate the growth time γ of an unstable longitudinal wave if the particle distribution function at $t = 0$ is

$$f_0(u_x) = A \exp\left[-\frac{(u_x - u_0)^2}{u_0^2}\right]$$

where u_0 is a constant such that $u_0 > v_p$.

BIBLIOGRAPHY

Bernstein, I. B., J. M. Greene, and M. D. Kruskal: Exact Nonlinear Plasma Oscillations, *Phys. Rev.*, **108**:546 (1957).
On a determination of the behavior of plasma waves which exist in the presence of trapping processes, using the Boltzmann equation.

Bohm, D., and E. P. Gross: Theory of Plasma Oscillations. *A*: Origin of Medium-like Behavior, *Phys. Rev.*, **75**:1851 (1949). Theory of Plasma Oscillations. *B*: Excitation and Damping of Oscillations, *Phys. Rev.*, **75**:1864 (1949).
A discussion of the physical mechanism for the growing and decaying of longitudinal plasma oscillations.

Dawson, J.: On Landau Damping, *Phys. Fluids*, **4**:869 (1961).
On a calculation of the dynamical details of the trapping process.

Jackson, J. D.: Plasma Oscillations, *J. Nucl. Energy, Pt. C, Plasma-Accelerators-Thermonucl. Res.*, **1**:171 (1960).
Includes a discussion of velocity-space instabilities similar to that given within the chapter.

Landau, L. D.: On the Vibrations of the Electronic Plasma, *J. Phys. (USSR)*, **9**:25 (1945).
On the calculation of Landau damping using the Boltzmann equation.

Spitzer, L., Jr.: "Physics of Fully Ionized Gases," 2d ed., Interscience Publishers, Inc., New York, 1962.
sec. 3.4*a*—Landau Damping. Includes a discussion of Landau damping similar to that given within the chapter.

Van Kampen, N. G.: On the Theory of Stationary Waves in a Plasma, *Physica*, **21**:949 (1955).
On a determination of the behavior of plasma waves which exist in the presence of trapping processes, using the Boltzmann equation.

appendix

A

The Derivation of the Momentum Transfer Equation and the Continuity Equation from the Boltzmann Equation

In Chap. 4 we derived the momentum transfer equation and the continuity equation for a given system of particles (electrons, ions, or neutrals) by considering that system of particles to behave as if it were a fluid whose velocity at any point was equal to the average velocity of the particles. In this appendix we shall justify the fluid approach by deriving the momentum transfer equation and the continuity equation from the Boltzmann equation. The Boltzmann equation for a system of particles with mass m and charge q can be written

$$\frac{\partial f}{\partial t} + \mathbf{u} \cdot \nabla_r f + \mathbf{a} \cdot \nabla_u f = \left(\frac{\partial f}{\partial t}\right)_{\text{collisions}} \tag{A-1}$$

where we shall consider only accelerations due to electric and magnetic fields (we shall neglect gravitational effects).

$$\mathbf{a} = \frac{q}{m}(\mathbf{E} + \mathbf{u} \times \mathbf{B}) \tag{A-2}$$

We shall first derive the continuity equation. If each term in the Boltzmann equation is integrated over all possible velocities, and if derivative and integral operations are interchanged, Eq. (A-1) becomes

$$\frac{\partial}{\partial t}\iiint_{\text{all }\mathbf{u}} f\, d^3u + \nabla_r \cdot \iiint_{\text{all }\mathbf{u}} \mathbf{u} f\, d^3u + \iiint_{\text{all }\mathbf{u}} a_x \frac{\partial f}{\partial u_x}\, d^3u + \iiint_{\text{all }\mathbf{u}} a_y \frac{\partial f}{\partial u_y}\, d^3u + \iiint_{\text{all }\mathbf{u}} a_z \frac{\partial f}{\partial u_z}\, d^3u = \iiint_{\text{all }\mathbf{u}} \left(\frac{\partial f}{\partial t}\right)_{\text{collisions}} d^3u \tag{A-3}$$

where the relationship $\mathbf{u} \cdot \nabla_r f = \nabla_r \cdot (\mathbf{u}f)$ has been employed. The first term on the left of Eq. (A-3) is the partial derivative with respect to time of the particle density. The next term on the left of Eq. (A-3) is the spatial gradient of the product of the particle density and the average velocity $\mathbf{v}$.

$$\frac{\partial n}{\partial t} + \nabla \cdot (n\mathbf{v}) = -\iiint_{\text{all } \mathbf{u}} a_x \frac{\partial f}{\partial u_x} d^3u - \iiint_{\text{all } \mathbf{u}} a_y \frac{\partial f}{\partial u_y} d^3u - \iiint_{\text{all } \mathbf{u}} a_z \frac{\partial f}{\partial u_z} d^3u + \iiint_{\text{all } \mathbf{u}} \left(\frac{\partial f}{\partial t}\right)_{\text{collisions}} d^3u \quad \text{(A-4)}$$

We shall now show that the four terms on the right of Eq. (A-4) are each equal to zero. Consider the first term on the right of Eq. (A-4). This term can be integrated by parts as follows:

$$-\iiint_{x_z u_y u_x} a_x \frac{\partial f}{\partial u_x} du_x\, du_y\, du_z = -\iint_{u_z u_y} [a_x f]_{-\infty}^{+\infty} du_y\, du_z + \iiint_{u_z u_y u_x} f \frac{\partial a_x}{\partial u_x} du_x\, du_y\, du_z \quad \text{(A-5)}$$

Since physical distribution functions go to zero at infinity more rapidly than accelerations become infinite, and since the acceleration in the x direction is independent of the velocity in the x direction, both terms on the right of Eq. (A-5) are equal to zero. It follows that each of the first three terms on the right of Eq. (A-4) is zero. The remaining term on the right of Eq. (A-4) represents the sum of the representative points scattered into all small volumes in phase space with the same real-space coordinates per unit time per unit real-space volume by collisions. If we assume that there is no creation or annihilation of particles, the same number of representative points must be scattered into some of these phase-space volumes as are scattered out of other phase-space volumes, since collisions are assumed not to change the position coordinates of the particles. The sum of the representative points scattered into the phase-space volumes under consideration is therefore zero. Equation (A-4) becomes the continuity equation

$$\frac{\partial n}{\partial t} + \nabla \cdot (n\mathbf{v}) = 0 \quad \text{(A-6)}$$

The technique for deriving the momentum transfer equation is similar to that used for deriving the continuity equation. We shall multiply each term in the Boltzmann equation by $m\mathbf{u}$ (instead of by 1) and then integrate each term over all velocities. After manipulating the resultant expression and making use of the continuity equation, we shall arrive at the momentum transfer equation. We first multiply Eq. (A-1) by $m\mathbf{u}$, integrate over all velocities, and interchange derivative and integral operations. Since

$$\mathbf{u}(\mathbf{u} \cdot \nabla_r f) = \nabla_r \cdot (\mathbf{uu}f)^*$$

we find that

$$\frac{\partial}{\partial t} \iiint\limits_{\text{all } \mathbf{u}} m\mathbf{u}f\, d^3u + \nabla_r \cdot \iiint\limits_{\text{all } \mathbf{u}} m\mathbf{uu}f\, d^3u + \iiint\limits_{\text{all } \mathbf{u}} (m\mathbf{u})(\mathbf{a} \cdot \nabla_u f)\, d^3u = \iiint\limits_{\text{all } \mathbf{u}} m\mathbf{u} \left(\frac{\partial f}{\partial t}\right)_{\text{collisions}} d^3u \quad \text{(A-7)}$$

where the product $\mathbf{uu}$ is a tensor quantity

$$\mathbf{uu} = \begin{bmatrix} u_x u_x & u_x u_y & u_x u_z \\ u_y u_x & u_y u_y & u_y u_z \\ u_z u_x & u_z u_y & u_z u_z \end{bmatrix}$$

The first term on the left of Eq. (A-7) can be written

$$\frac{\partial}{\partial t} \iiint\limits_{\text{all } \mathbf{u}} m\mathbf{u}f\, d^3u = \frac{\partial}{\partial t}(nm\mathbf{v}) \quad \text{(A-8)}$$

The second term on the left of Eq. (A-7) can be written

$$\nabla_r \cdot \iiint\limits_{\text{all } \mathbf{u}} m\mathbf{uu}f\, d^3u = \nabla_r \cdot (nm)\overline{\mathbf{uu}} \quad \text{(A-9)}$$

The third term on the left of Eq. (A-7) can be integrated by parts with the result that

$$\iiint\limits_{\text{all } \mathbf{u}} (m\mathbf{u})(\mathbf{a} \cdot \nabla_u f)\, d^3u = -nm\bar{\mathbf{a}} \quad \text{(A-10)}$$

Thus Eq. (A-7) becomes

$$\frac{\partial}{\partial t}(nm\mathbf{v}) + \nabla_r \cdot (nm\overline{\mathbf{uu}}) - nm\bar{\mathbf{a}} = \iiint\limits_{\text{all } \mathbf{u}} m\mathbf{u} \left(\frac{\partial f}{\partial t}\right)_{\text{collisions}} d^3u \quad \text{(A-11)}$$

* This identity may be proved by making use of Eqs. (D-4) and (D-9) and of the definition of the tensor quantity $\mathbf{uu}$.

We shall make the substitution $\mathbf{u} = \mathbf{v} + \mathbf{w}$, where $\mathbf{v}$ is the average velocity and $\mathbf{w}$ is the remaining component of velocity, into the second term on the left-hand side of Eq. (A-11).

$$\nabla_r \cdot (nm\overline{\mathbf{uu}}) = \nabla_r \cdot (nm\mathbf{vv}) + \nabla_r \cdot (nm\overline{\mathbf{ww}}) + \nabla_r \cdot (nm\overline{\mathbf{w}}\mathbf{v}) + \nabla_r \cdot (nm\mathbf{v}\overline{\mathbf{w}}) \qquad \text{(A-12)}$$

We have not averaged over $\mathbf{v}$ in Eq. (A-12) because $\mathbf{v}$ is already an average quantity. The average value of $\mathbf{w}$ is zero.[1] If the first term on the right of Eq. (A-12) is expanded, Eq. (A-12) can be written

$$\nabla_r \cdot (nm\overline{\mathbf{uu}}) = nm(\mathbf{v} \cdot \nabla_r)\mathbf{v} + \mathbf{v}\nabla_r \cdot (nm\mathbf{v}) + \nabla_r \cdot (nm\overline{\mathbf{ww}}) \qquad \text{(A-13)}$$

The first term on the left of Eq. (A-11) can be expanded as

$$\frac{\partial}{\partial t}(nm\mathbf{v}) = nm\frac{\partial \mathbf{v}}{\partial t} + \mathbf{v}\frac{\partial (nm)}{\partial t} \qquad \text{(A-14)}$$

Equations (A-2), (A-13), and (A-14) can be substituted into Eq. (A-11) with the result that

$$nm\frac{\partial \mathbf{v}}{\partial t} + \mathbf{v}\frac{\partial (nm)}{\partial t} + nm(\mathbf{v} \cdot \nabla_r)\mathbf{v} + \mathbf{v}\nabla_r \cdot (nm\mathbf{v}) + \nabla_r \cdot (nm\overline{\mathbf{ww}}) - nq(\mathbf{E} + \mathbf{v} \times \mathbf{B}) = \iiint_{\text{all } \mathbf{u}} m\mathbf{u}\left(\frac{\partial f}{\partial t}\right)_{\text{collisions}} d^3u \qquad \text{(A-15)}$$

Equation (A-6), the continuity equation, can be written

$$\frac{\partial n}{\partial t} = -\nabla \cdot (n\mathbf{v})$$

If $\frac{\partial n}{\partial t}$ from the continuity equation is substituted into the second term on the left side of Eq. (A-15), the resultant equation is

$$nm\frac{\partial \mathbf{v}}{\partial t} + nm(\mathbf{v} \cdot \nabla_r)\mathbf{v} + \nabla_r \cdot (nm\overline{\mathbf{ww}}) - nq(\mathbf{E} + \mathbf{v} \times \mathbf{B}) = \iiint_{\text{all } \mathbf{u}} m\mathbf{u}\left(\frac{\partial f}{\partial t}\right)_{\text{collisions}} d^3u$$

---

[1] $$\overline{\mathbf{w}} = \frac{\iiint_{\text{all } \mathbf{u}} (\mathbf{u} - \mathbf{v}) f\, d^3u}{\iiint_{\text{all } \mathbf{u}} f\, d^3u} = \frac{\iiint_{\text{all } \mathbf{u}} \mathbf{u} f\, d^3u}{\iiint_{\text{all } \mathbf{u}} f\, d^3u} - \frac{\mathbf{v}\iiint_{\text{all } \mathbf{u}} f\, d^3u}{\iiint_{\text{all } \mathbf{u}} f\, d^3u} = \mathbf{v} - \mathbf{v} = 0$$

or

$$nm\frac{d\mathbf{v}}{dt} = -\nabla_r \cdot (nm\overline{\mathbf{ww}}) + nq(\mathbf{E} + \mathbf{v} \times \mathbf{B}) + \iiint_{\text{all }\mathbf{u}} m\mathbf{u}\left(\frac{\partial f}{\partial t}\right)_{\text{collisions}} d^3u \tag{A-16}$$

where

$$nm\left[\frac{\partial \mathbf{v}}{dt} + (\mathbf{v} \cdot \nabla_r)\mathbf{v}\right] = nm\frac{d\mathbf{v}}{dt}$$

The last term on the right of Eq. (A-16) is the total momentum gained per unit time per unit real-space volume by the particles described by the distribution function as a result of collisions. Collisions of these identical particles with each other produce no net momentum gain. Collisions of these particles with other particles may yield a net momentum change, so that the last term on the right of Eq. (A-16) represents the momentum change per unit time per unit real-space volume of the particles we are considering due to collisions with other particles. We shall assume that it is possible to write

$$\iiint_{\text{all }\mathbf{u}} m\mathbf{u}\left(\frac{\partial f}{\partial t}\right)_{\text{collisions}} d^3u = -nm\nu_c\mathbf{v} \tag{A-17}$$

where Eq. (A-17) can be considered a definition of the effective collision frequency ν_c.*

The first term on the right of Eq. (A-16) can be written[1]

$$\nabla_r \cdot (nm\overline{\mathbf{ww}}) = \begin{bmatrix}\dfrac{\partial}{\partial x} & \dfrac{\partial}{\partial y} & \dfrac{\partial}{\partial z}\end{bmatrix}\begin{bmatrix} nm\overline{w_xw_x} & nm\overline{w_xw_y} & nm\overline{w_xw_z} \\ nm\overline{w_yw_x} & nm\overline{w_yw_y} & nm\overline{w_yw_z} \\ nm\overline{w_zw_x} & nm\overline{w_zw_y} & nm\overline{w_zw_z}\end{bmatrix} \tag{A-18}$$

We shall assume that the distribution function is isotropic in $\mathbf{w}$. Thus $\overline{w_xw_y} = \overline{w_xw_z} = \overline{w_yw_z} = 0$ and $\overline{w_x^2} = \overline{w_y^2} = \overline{w_z^2} = \overline{w^2}/3$. Equation (A-18) therefore becomes

$$\nabla_r \cdot (nm\overline{\mathbf{ww}}) = \nabla_r\left(\frac{nm\overline{w^2}}{3}\right) \tag{A-19}$$

The term $(nm\overline{w^2}/3)$ is the kinetic pressure P derived in Chap. 3. If Eqs. (A-17) and (A-19) are substituted into Eq. (A-16), the resulting expression is the momentum transfer equation which we derived by other techniques in Chap. 4.

$$\frac{d\mathbf{v}}{dt} = -\frac{\nabla P}{mn} + \frac{q}{m}(\mathbf{E} + \mathbf{v} \times \mathbf{B}) - \nu_c\mathbf{v} \tag{A-20}$$

* See Appendix B for an alternative form of the collision term.

[1] See Eqs. (D-8) and (D-9).

It is worthwhile noting that additional hydrodynamic equations may be derived by multiplying the Boltzmann equation by higher-order terms in $\mathbf{u}$ and then integrating over all velocity coordinates. In some instances these higher-order "transfer equations" are very useful.

BIBLIOGRAPHY

Delcroix, J. L.: "Introduction to the Theory of Ionized Gases," chap. 8, Interscience Publishers, Inc., New York, 1960.

Jeans, J. H.: "The Dynamical Theory of Gases," 4th ed., chap. ix, Dover Publications, Inc., New York, 1954.

Spitzer, L., Jr.: "Physics of Fully Ionized Gases," 2d ed., appendix, Interscience Publishers, Inc., New York, 1962.

appendix

B

The Frictional-force Term and the Effective Collision Frequency

In Chap. 4 we defined *effective* collision frequencies in order to take account of the frictional forces which the three intermingled plasma fluids exert on one another. The exact frictional-force terms to be used in the momentum transfer equations can be determined only by deriving the momentum transfer equations from the appropriate Boltzmann equations. We use the frictional-force terms defined in Chap. 4 throughout this text for two reasons: (1) The assumed frictional-force terms lead to momentum transfer equations which can be mathematically manipulated with relative ease, and (2) the assumed frictional-force terms are useful approximations to the exact frictional-force terms in many cases (particularly in plasmas with a low degree of ionization). In this appendix we shall examine an alternative form for the frictional force terms.

We shall first reiterate those arguments which lead to the definition of an *effective* collision frequency. We shall consider the fluid of plasma electrons. The results of our discussion can be easily extended to include the ion fluid and the neutral-particle fluid. We assumed in Chap. 4 that the frictional force on the electron fluid due to electron-ion collisions is proportional to the density of the electrons, to the density of the ions, and to the velocity of the electron fluid and that the frictional force on the electron fluid due to electron-neutral collisions is proportional to the density of the electrons, to the density of the neutrals, and to the velocity of the electron fluid. Thus the frictional force on the electron fluid contained within the differential volume d^3r could be written

$$dF_f = -n_e(\alpha_{ei}n_i + \alpha_{en}n_n)\mathbf{v}_e\, d^3r \tag{4-8}$$

where n_e, n_i, n_n are functions of position coordinates and the α's are variable factors to take complete account of the collision interactions. The term α_{ei} may be a function of $\mathbf{v}_e$ and $\mathbf{v}_i$; the term α_{en} may be a function of $\mathbf{v}_e$ and $\mathbf{v}_n$. The substitution of Eq. (4-8) into Newton's equation of motion for the small volume of electron fluid leads to the following definition of the electron effective collision frequency:

$$\nu_{ce} \equiv \frac{(\alpha_{ei}n_i + \alpha_{en}n_n)}{m_e} \tag{B-1}$$

It follows that the effective collision frequencies for the ion fluid and for the neutral-particle fluid are, respectively,

$$\nu_{ci} \equiv \frac{(\alpha_{ie}n_e + \alpha_{in}n_n)}{m_i} \tag{B-2}$$

$$\nu_{cn} \equiv \frac{(\alpha_{ni}n_i + \alpha_{ne}n_e)}{m_n} \tag{B-3}$$

It should be noted that

$$\alpha_{ei} \neq \alpha_{ie} \tag{B-4}$$

$$\alpha_{en} \neq \alpha_{ne} \tag{B-5}$$

$$\alpha_{ni} \neq \alpha_{in} \tag{B-6}$$

We shall now express the frictional-force terms in an alternative form. We assume that the frictional force between two fluids is proportional to the densities of the two fluids and to the *difference* in the velocities of the two fluids. Thus we can write the electron frictional-force term as

$$dF_f = -n_e[n_i\eta_{ei}(\mathbf{v}_e - \mathbf{v}_i) + n_n\eta_{en}(\mathbf{v}_e - \mathbf{v}_n)]\, d^3r \tag{B-7}$$

where the η's are factors to take complete account of the collision interaction. The η's are assumed not to be functions of the $\mathbf{v}$'s. From a comparison of Eqs. (4-8), (B-1), and (B-7), we find that

$$\nu_{ce}\mathbf{v}_e \rightarrow \frac{[n_i\eta_{ei}(\mathbf{v}_e - \mathbf{v}_i) + n_n\eta_{en}(\mathbf{v}_e - \mathbf{v}_n)]}{m_e} \tag{B-8}$$

If we use the type of frictional-force terms specified by Eq. (B-7), the three momentum transfer equations can be written

$$\frac{d\mathbf{v}_e}{dt} = -\frac{\nabla P_e}{n_e m_e} - \frac{|e|}{m_e}(\mathbf{E} + \mathbf{v}_e \times \mathbf{B}) - \frac{[n_i\eta_{ei}(\mathbf{v}_e - \mathbf{v}_i) + n_n\eta_{en}(\mathbf{v}_e - \mathbf{v}_n)]}{m_e} \tag{B-9}$$

$$\frac{d\mathbf{v}_i}{dt} = -\frac{\nabla P_i}{n_i m_i} + \frac{Z_i|e|}{m_i}(\mathbf{E} + \mathbf{v}_i \times \mathbf{B}) - \frac{[n_e\eta_{ie}(\mathbf{v}_i - \mathbf{v}_e) + n_n\eta_{in}(\mathbf{v}_i - \mathbf{v}_n)]}{m_i} \tag{B-10}$$

$$\frac{d\mathbf{v}_n}{dt} = -\frac{\nabla P_n}{n_n m_n} - \frac{[n_i\eta_{ni}(\mathbf{v}_n - \mathbf{v}_i) + n_e\eta_{ne}(\mathbf{v}_n - \mathbf{v}_e)]}{m_n} \tag{B-11}$$

Whichever set of momentum transfer equations is used, the three momentum transfer equations are coupled by their frictional-force terms. In the set of momentum transfer equations given above, this coupling is the more explicit. In addition, the frictional-force terms given above are more general than those used throughout the text in that they allow the frictional force on a fluid of velocity $\mathbf{v}(\mathbf{r})$ to be in a direction other than $\mathbf{v}$.

Since the frictional force which the electron fluid exerts on the ion fluid is equal and opposite to the frictional force which the ion fluid exerts on the electron fluid, it follows from an examination of the momentum transfer equations given above that

$$\eta_{ei} = \eta_{ie} \tag{B-12}$$

A similar argument can be advanced to show that

$$\eta_{en} = \eta_{ne} \tag{B-13}$$

$$\eta_{ni} = \eta_{in} \tag{B-14}$$

The formulation of the momentum transfer equations given in Eqs. (B-9), (B-10), and (B-11) is particularly useful for investigating the properties of a fully ionized plasma, that is, $n_n = 0$. For the case of a fully ionized plasma, only two coupled equations [Eqs. (B-9) and (B-10) with $n_n = 0$] need be considered. These two equations can be solved in conjunction with the continuity and state equations to determine the behavior of the fully ionized plasma under various conditions.

BIBLIOGRAPHY

Spitzer, L., Jr.: "Physics of Fully Ionized Gases," 2d ed., Interscience Publishers, Inc., New York, 1962.
p. 28—A discussion of the assumed form of the frictional-force term.

appendix

C

Vector Relations

C-1 Vector Identities

$$
\begin{aligned}
\mathbf{A}\cdot\mathbf{B} &= \mathbf{B}\cdot\mathbf{A}\\
\mathbf{A}\times\mathbf{B} &= -\mathbf{B}\times\mathbf{A}\\
\mathbf{A}\times(\mathbf{B}\times\mathbf{C}) &= (\mathbf{A}\cdot\mathbf{C})\mathbf{B}-(\mathbf{A}\cdot\mathbf{B})\mathbf{C}\\
(\mathbf{A}\times\mathbf{B})\times\mathbf{C} &= (\mathbf{C}\cdot\mathbf{A})\mathbf{B}-(\mathbf{B}\cdot\mathbf{C})\mathbf{A}\\
\mathbf{A}\cdot(\mathbf{B}\times\mathbf{C}) &= \mathbf{B}\cdot(\mathbf{C}\times\mathbf{A}) = \mathbf{C}\cdot(\mathbf{A}\times\mathbf{B})\\
\nabla(\phi+\Psi) &= \nabla\phi+\nabla\Psi\\
\nabla(\phi\Psi) &= \phi\nabla\Psi+\Psi\nabla\phi\\
\nabla\cdot(\phi\mathbf{A}) &= \phi\nabla\cdot\mathbf{A}+\mathbf{A}\cdot\nabla\phi\\
\nabla\times(\phi\mathbf{A}) &= \phi(\nabla\times\mathbf{A})+(\nabla\phi)\times\mathbf{A}\\
\nabla\cdot(\mathbf{A}\times\mathbf{B}) &= \mathbf{B}\cdot(\nabla\times\mathbf{A})-\mathbf{A}\cdot(\nabla\times\mathbf{B})\\
\nabla\times(\mathbf{A}\times\mathbf{B}) &= (\mathbf{B}\cdot\nabla)\mathbf{A}-(\mathbf{A}\cdot\nabla)\mathbf{B}+\mathbf{A}(\nabla\cdot\mathbf{B})-\mathbf{B}(\nabla\cdot\mathbf{A})\\
\nabla(\mathbf{A}\cdot\mathbf{B}) &= (\mathbf{A}\cdot\nabla)\mathbf{B}+(\mathbf{B}\cdot\nabla)\mathbf{A}+\mathbf{A}\times(\nabla\times\mathbf{B})+\mathbf{B}\times(\nabla\times\mathbf{A})\\
\nabla\times\nabla\phi &= 0\\
\nabla\cdot\nabla\times\mathbf{A} &= 0\\
\nabla\times(\nabla\times\mathbf{A}) &= \nabla(\nabla\cdot\mathbf{A})-(\nabla\cdot\nabla)\mathbf{A}
\end{aligned}
$$

$$\iiint\limits_{\text{vol}} (\nabla\cdot\mathbf{A})\,d^3r = \oint\!\!\!\oint\limits_{\substack{\text{surface}\\ \text{enclosing}\\ \text{vol}}} \mathbf{A}\cdot\mathbf{n}\,dA$$

$$\iint\limits_{\text{area}} (\nabla\times\mathbf{A})\cdot\mathbf{n}\,dA = \oint\limits_{\substack{\text{circumference}\\ \text{of area}}} \mathbf{A}\cdot d\mathbf{L}$$

C-2 The Vector Quantity $(\mathbf{B}\cdot\nabla)\mathbf{A}$

In cartesian coordinates the quantity $(\mathbf{B}\cdot\nabla)\mathbf{A}$ is defined as

$$(\mathbf{B}\cdot\nabla)\mathbf{A} \equiv B_x\frac{\partial\mathbf{A}}{\partial x}+B_y\frac{\partial\mathbf{A}}{\partial y}+B_z\frac{\partial\mathbf{A}}{\partial z}$$

Thus, in component form,

$$(\mathbf{B}\cdot\nabla)\mathbf{A} = \left(B_x\frac{\partial A_x}{\partial x} + B_y\frac{\partial A_x}{\partial y} + B_z\frac{\partial A_x}{\partial z}\right)\mathbf{a}_x + \left(B_x\frac{\partial A_y}{\partial x} + B_y\frac{\partial A_y}{\partial y} + B_z\frac{\partial A_y}{\partial z}\right)\mathbf{a}_y + \left(B_x\frac{\partial A_z}{\partial x} + B_y\frac{\partial A_z}{\partial y} + B_z\frac{\partial A_z}{\partial z}\right)\mathbf{a}_z$$

In cylindrical coordinates, the quantity $(\mathbf{B}\cdot\nabla)\mathbf{A}$ is defined as

$$(\mathbf{B}\cdot\nabla)\mathbf{A} \equiv B_r\frac{\partial \mathbf{A}}{\partial r} + \frac{B_\theta}{r}\frac{\partial \mathbf{A}}{\partial \theta} + B_z\frac{\partial \mathbf{A}}{\partial z}$$

The cylindrical coordinate unit vectors $\mathbf{a}_\theta$, $\mathbf{a}_r$ are functions of the θ coordinate

$$\frac{\partial \mathbf{a}_\theta}{\partial \theta} = -\mathbf{a}_r$$

$$\frac{\partial \mathbf{a}_r}{\partial \theta} = \mathbf{a}_\theta$$

It follows that in component form,

$$(\mathbf{B}\cdot\nabla)\mathbf{A} = \left(B_r\frac{\partial A_r}{\partial r} + \frac{B_\theta}{r}\frac{\partial A_r}{\partial \theta} + B_z\frac{\partial A_r}{\partial z} - \frac{A_\theta B_\theta}{r}\right)\mathbf{a}_r + \left(B_r\frac{\partial A_\theta}{\partial r} + \frac{B_\theta}{r}\frac{\partial A_\theta}{\partial \theta} + B_z\frac{\partial A_\theta}{\partial z} + \frac{A_r B_\theta}{r}\right)\mathbf{a}_\theta + \left(B_r\frac{\partial A_z}{\partial r} + \frac{B_\theta}{r}\frac{\partial A_z}{\partial \theta} + B_z\frac{\partial A_z}{\partial z}\right)\mathbf{a}_z$$

appendix

D

Elementary Tensor Operations

We shall consider a three-dimensional coordinate system in which are defined three orthogonal unit vectors $\mathbf{a}_1$, $\mathbf{a}_2$, $\mathbf{a}_3$. Any vector can be expressed in terms of the three unit vectors. For example,

$$\mathbf{A} = A_1\mathbf{a}_1 + A_2\mathbf{a}_2 + A_3\mathbf{a}_3 \tag{D-1}$$

We define a second-order tensor in this coordinate system as a quantity having nine elements and written as follows:

$$\langle\gamma\rangle = \begin{bmatrix} \gamma_{11} & \gamma_{12} & \gamma_{13} \\ \gamma_{21} & \gamma_{22} & \gamma_{23} \\ \gamma_{31} & \gamma_{32} & \gamma_{33} \end{bmatrix} \tag{D-2}$$

We shall discuss only second-order tensors. We shall define the following tensor operations: (1) addition of second-order tensors, (2) multiplication of a second-order tensor by a scalar, (3) multiplication of a second-order tensor and a vector, and (4) multiplication of a vector and a second-order tensor.

1. The addition of the two tensors $\langle\gamma\rangle$ and $\langle\beta\rangle$ is defined as the arithmetic addition of the elements of those tensors.

$$\langle\gamma\rangle + \langle\beta\rangle \equiv \begin{bmatrix} \gamma_{11}+\beta_{11} & \gamma_{12}+\beta_{12} & \gamma_{13}+\beta_{13} \\ \gamma_{21}+\beta_{21} & \gamma_{22}+\beta_{22} & \gamma_{23}+\beta_{23} \\ \gamma_{31}+\beta_{31} & \gamma_{32}+\beta_{32} & \gamma_{33}+\beta_{33} \end{bmatrix} \tag{D-3}$$

2. The multiplication of a tensor by a scalar is defined as the arithmetic multiplication of each of the elements of the tensor by the scalar. Thus the product of the scalar h and the tensor $\langle\gamma\rangle$ is written

$$h\langle\gamma\rangle \equiv \begin{bmatrix} h\gamma_{11} & h\gamma_{12} & h\gamma_{13} \\ h\gamma_{21} & h\gamma_{22} & h\gamma_{23} \\ h\gamma_{31} & h\gamma_{32} & h\gamma_{33} \end{bmatrix} \tag{D-4}$$

3. The product of the tensor $\langle\gamma\rangle$ and the vector $\mathbf{A}$ is written

$$\langle\gamma\rangle\cdot\mathbf{A} = \begin{bmatrix} \gamma_{11} & \gamma_{12} & \gamma_{13} \\ \gamma_{21} & \gamma_{22} & \gamma_{23} \\ \gamma_{31} & \gamma_{32} & \gamma_{33} \end{bmatrix}\begin{bmatrix} A_1 \\ A_2 \\ A_3 \end{bmatrix} \tag{D-5}$$

By definition, the ith component of $\langle\gamma\rangle\cdot\mathbf{A}$ is

$$(\langle\gamma\rangle\cdot\mathbf{A})_i \equiv \sum_j \gamma_{ij}A_j \tag{D-6}$$

where i and j each assume the values 1, 2, 3. It follows that

$$\langle\gamma\rangle\cdot\mathbf{A} = (\gamma_{11}A_1 + \gamma_{12}A_2 + \gamma_{13}A_3)\mathbf{a}_1 + (\gamma_{21}A_1 + \gamma_{22}A_2 + \gamma_{23}A_3)\mathbf{a}_2 + (\gamma_{31}A_1 + \gamma_{32}A_2 + \gamma_{33}A_3)\mathbf{a}_3 \tag{D-7}$$

4. The product of the vector $\mathbf{A}$ and the tensor $\langle\gamma\rangle$ is written

$$\mathbf{A}\cdot\langle\gamma\rangle = [A_1 \quad A_2 \quad A_3]\begin{bmatrix} \gamma_{11} & \gamma_{12} & \gamma_{13} \\ \gamma_{21} & \gamma_{22} & \gamma_{23} \\ \gamma_{31} & \gamma_{32} & \gamma_{33} \end{bmatrix} \tag{D-8}$$

By definition, the ith component of $\mathbf{A}\cdot\langle\gamma\rangle$ is

$$(\mathbf{A}\cdot\langle\gamma\rangle)_i \equiv \sum_j A_j\gamma_{ji} \tag{D-9}$$

where i and j each assume the values 1, 2, 3. It follows that

$$\mathbf{A}\cdot\langle\gamma\rangle = (A_1\gamma_{11} + A_2\gamma_{21} + A_3\gamma_{31})\mathbf{a}_1 + (A_1\gamma_{12} + A_2\gamma_{22} + A_3\gamma_{32})\mathbf{a}_2 + (A_1\gamma_{13} + A_2\gamma_{23} + A_3\gamma_{33})\mathbf{a}_3 \tag{D-10}$$

appendix

E

Physical Constants

c	Speed of light in vacuum	2.99793×10^{8} m/sec
ϵ_0	Permittivity of vacuum	8.854×10^{-12} farad/m
μ_0	Permeability of vacuum	1.257×10^{-6} henry/m
κ	Boltzmann constant	1.3804×10^{-23} joule/°K
$\|e\|$	Charge of proton	1.6021×10^{-19} coul
m_e	Rest mass of electron	9.108×10^{-31} kg
m_p	Rest mass of proton	1.672×10^{-27} kg

Index